Science

YEAR 3

Sue Hunter
Jenny Macdonald

GALORE PARK

AN HACHETTE UK COMPANY

The Publishers would like to thank the following for permission to reproduce copyright material.

Photo credits

page ix © Pressmaster/Fotolia **p1** © margostock/Fotolia **p4 (centre)** © cube197/Fotolia **p4 (bottom)** © blickwinkel / Alamy **p5** © Goldbany/Fotolia **p8** © Andreanita/Fotolia **p13** © POWER AND SYRED/SCIENCE PHOTO LIBRARY **p15 (top)** © bdspn/ iStockphoto/ Thinkstock **p15 (bottom)** © pr2is/Fotolia.com **p19** © Kostia Lomzov/Fotolia **p20** © jurra8/Fotolia **p21 (top)** Brian Jackson/Fotolia © **p21 (centre)** © WaterFrame/Alamy **p21 (bottom)** © CHARLES E MOHR/SCIENCE PHOTO LIBRARY **p23 (left)** © ALEX HYDE/ SCIENCE PHOTO LIBRARY **p23 (right)** © DR. JOHN BRACKENBURY/SCIENCE PHOTO LIBRARY **p24 (left)** © oliver leedham/Alamy **p24 (right)** © M. Schuppich/Fotolia **p25** © GUSTOIMAGES/SCIENCE PHOTO LIBRARY **p27** © Sue Hunter **p32** © ia_64/Fotolia **p33 (top)** © Image Source/Alamy **p33 (bottom)** © Viktor/Fotolia **p34 (top)** © Dethchimo/Fotolia **p34 (centre)** © Africa Studio/Fotolia **p34 (bottom)** © BIOPHOTO ASSOCIATES/SCIENCE PHOTO LIBRARY **p35 (top)** © ematon **p35 (bottom)** DAVID MUNNS/SCIENCE PHOTO LIBRARY **p39** © vetasster/Fotolia **p40** © GL Archive/Alamy **p41 (left)** © Karen Hadley/Fotolia **p41 (right)** © Steve Byland/ Fotolia **p44** © Friday/ Fotolia **p46 (top)** © Alex Wilson/Getty Images **p47 (top)** © Jamenpercy/Fotolia **p47 (centre)** © Vinicius Tupinamba/Fotolia **p50** © TOM MCHUGH/SCIENCE PHOTO LIBRARY **p52** © Beboy/ Fotolia **p56** © Gekaskr/Fotolia **p58 (top)** © Santi Rodríguez/Fotolia **p58 (bottom)** © hal_pand_108/ Fotolia **p61 (top)** © Smereka/ Fotolia **p61 (bottom)** © Sue Hunter **p62, p63** © Springfield Gallery/Fotolia **p64, p72, p73** © Sue Hunter **p79** © Sergey Nivens/ Fotolia **p81** © Karelnoppe/Thinkstock **p82** © Tinx/Fotolia **p87 (top)** © Anrymos/Fotolia **p87 (bottom)** © Les Cunliffe/Thinkstock **p88** © Peshkova/Fotolia **p90** © Katarinagondova/Fotolia **p91** © Tim Large - Medical themed/Alamy **p93 (left)** © sniper815/Fotolia **p93 (right)** © ALFRED PASIEKA/SCIENCE PHOTO LIBRARY **p96** © andrea lehmkuhl/Fotolia **p100 (top)** © Editorial Image, LLC/Alamy **p101** © Fovivafoto/Fotolia **p102 (top)** © Ed Brown/Alamy **p102 (bottom)** © Uwimages/ Fotolia **p105** © Jackie Barns-Graham **p113** © Shuke/Fotolia **p114** © Studio49/Fotolia

Every effort has been made to trace all copyright holders, but if any have been inadvertently overlooked, the Publishers will be pleased to make the necessary arrangements at the first opportunity.

Although every effort has been made to ensure that website addresses are correct at time of going to press, Galore Park cannot be held responsible for the content of any website mentioned in this book. It is sometimes possible to find a relocated web page by typing in the address of the home page for a website in the URL window of your browser.

Hachette UK's policy is to use papers that are natural, renewable and recyclable products and made from wood grown in sustainable forests. The logging and manufacturing processes are expected to conform to the environmental regulations of the country of origin.

Orders: please contact Hachette UK Distribution, Hely Hutchinson Centre, Milton Road, Didcot, Oxfordshire, OX11 7HH. Telephone: +44 (0)1235 827827. Email education@hachette.co.uk. Lines are open from 9 a.m. to 5 p.m., Monday to Friday. You can also order through our website: www.hoddereducation.co.uk. If you have queries or questions that aren't about an order, you can contact us at hoddergibson@hodder.co.uk

Parents, Tutors please call: 020 3122 6405 (Monday to Friday, 9:30 a.m. to 4.30 p.m.). Email: parentenquiries@galorepark.co.uk

Visit our website at www.galorepark.co.uk for details of other revision guides for Common Entrance, examination papers and Galore Park publications.

ISBN: 978 1 4718 5628 0

© Sue Hunter and Jenny Macdonald 2015

First published in 2015 by Galore Park Publishing Limited
An Hachette UK Company
Carmelite House
50 Victoria Embankment
London EC4Y 0DZ
www.galorepark.co.uk

Impression number   10 9

Year   2023

Cover photo © Chepko Danil - Fotolia.com
Illustrations by Aptara, Inc.
Typeset in India by Aptara, Inc.
Printed by CPI Group (UK) Ltd, Croydon CR0 4YY

A catalogue record for this title is available from the British Library.

About the authors

Sue Hunter has very recently retired as full-time science teacher but she continues to be very involved in Science education. She is a member of the Common Entrance setting team and a governor of local primary schools. Sue has written extensively, including Galore Park's KS2 Science textbooks and 11+ Science Revision Guide.

Jenny Macdonald has had a happy and fulfilling career as a teacher, teaching in both state and private schools. She has recently retired having spent the last eighteen years teaching science in a local prep school. In the last few years she has contributed to Galore Park's KS2 Science textbooks.

Preface

The universe is full of magical things patiently waiting for our wits to grow sharper. **Eden Phillpotts**

The study of science for young children is a voyage of discovery. It stimulates their curiosity and provides a vehicle for them to explore their world, to ask questions about things that they observe and to make sense of their observations. It does not exist in isolation but draws upon many other aspects of a well-rounded curriculum and should be practical, interesting and, above all, fun.

This book covers the requirements for the National Curriculum for Year 3. It also contains additional material as necessary to meet the specification for Year 3 in the ISEB Common Entrance syllabus and some extension material. It includes ideas for activities to develop practical skills, deepen understanding and provide stimulus for discussion and questioning.

Practical work is always popular and hands-on activities in this book are designed to be carried out by the pupils in pairs or small groups. Pupils should be encouraged to think about safety at all times when carrying out practical activities. However, the responsibility for risk assessment lies with the teacher, who should ideally try out each activity before presenting it to the class in order to identify any risks as appropriate to the particular group of children involved. The ASE publication *Be safe!* (available via the ASE website: www.ase.org.uk/resources) is a useful source of information and advice about risk assessment in the primary phase.

Exercises have been set at intervals throughout the book. Where there is more than one exercise in a group, the first one is set at standard level followed by a more easily accessible exercise covering the same material and/or an extension exercise.

Notes on features

Words printed in **blue and bold** are keywords. All keywords are defined in the glossary near the end of the book.

Exercise

Exercises of varying lengths are provided to give you plenty of opportunities to practise what you have learnt.

Activity

Sometimes it is useful to explore a topic in more detail by researching it. An activity is an opportunity to discover interesting things for yourself, and to practise recording and presenting what you find out. Some activities provide opportunities for you to do experiments. Others need some research from books or the internet, or maybe by talking to other people.

Did you know?

In these boxes you will learn interesting and often surprising facts about the natural world to inform your understanding of each topic.

Working Scientifically is an important part of learning science. When you see this mark you will be practising the really important skills that make good scientists. You will find out:

Working Scientifically

- why we carry out experiments
- what we mean by the word 'variable'
- what we mean by a fair test
- how to design experiments to answer your own questions

- how to measure variables

- how to record and display results clearly and accurately

- how scientific understanding is built up by the work of many scientists learning from each other, sometimes over hundreds of years.

Go further

The material in these boxes goes beyond the ISEB syllabus for 11+. You do not need to learn it for an 11+ exam but your teacher may decide that it is a good idea for you to learn something a bit extra to help you to understand a topic better or to extend your learning. All this material will be useful to you in your future studies ...

Contents

	Introduction	viii
1	Life processes	1
2	Green plants	9
3	A healthy diet	30
4	Skeleton and movement	43
5	Rocks	51
6	Soils	61
7	Light	76
8	Friction and movement	94
9	Magnets	107
	Glossary	119
	Index	123

Introduction

➡ How to be a scientist

What is science?

The word 'science' comes from the Latin word for knowledge, so science is knowledge. However, we usually think of science as being the study of the world around us.

There are lots of different things to study in our world. Through science we can find out about plants and animals that live on the planet Earth. We can find out about where they live, how they feed and breed and how they all depend on each other. We can learn about how our bodies work and what we can do to keep ourselves fit and healthy.

We can find out about the rocks and soils that make up the Earth. From these rocks we can get materials to use in our everyday lives. We need to learn about these materials, how they behave and how they can be used to make the things we need. This will help us find new materials or make better use of the ones we already have.

It is very exciting to investigate the way things work. We can learn about the forces that make things move and the energy that drives them. We can look up and study the planets, stars and galaxies that make up the Universe.

When you study science at school, you will start by learning about things that other people have discovered. But anyone can find out something new that no one has ever thought of before. The important thing is to take an interest in everything that is going on around you.

You need to think about what you see and hear and to ask questions about anything that puzzles you. The world is a huge and complicated place. We don't know everything about it so there are lots of new things to discover.

■ Science helps us to answer questions about the world

Activity – what do scientists do?

There are many different types of scientist who study many different things. They are often given special names, which give us a clue about the sort of things they investigate. See if you can find out what would be studied by each of the following kinds of scientist:

- astronomer
- biologist
- botanist
- entomologist

- geologist
- herpetologist
- palaeontologist
- physicist

Working like a scientist

Scientists find out about the world around them in lots of ways. To be a good scientist you need to:

- observe
- record
- ask questions

- find answers
- work safely.

Observe

Every day you take in lots of information about the world around you. Your five senses (sight, hearing, smell, taste and touch) are busy all the time. When you notice something interesting, you can look more closely or listen more carefully to find out more. This is observing, and good scientists do it all the time.

Record

Scientists keep records of their observations. Records could be notes, photographs, diagrams, measurements, or video or sound recordings. Recording the things they have observed helps them to remember the interesting facts they have discovered. They can use their records to help them to find patterns in their observations so that they can explain what they observe.

Ask questions

Observation often leads to questions. Questions lead you to investigate further. Asking questions is one of the best ways of learning more about the world around you. Often your parents or teachers can answer your questions. Sometimes they cannot so you will need to find out for yourself.

Find answers

There are many ways to find answers to your questions. You can observe more closely or look in books or on the internet. Most scientists will find the answers to their questions by doing experiments. You will be doing lots of experiments during your study of science.

You may not realise it, but you have been a scientist all your life! Even very small babies observe the world around them, and young children ask questions all the time. Your science lessons will help you to become an even better scientist.

Work safely

You may have your science lessons in your classroom or you may go to a special science room, called a laboratory. Wherever you are studying, it is very important that you work safely at all times, especially when doing experiments. Your school will have a special set of rules that you must follow when doing science experiments and it is important to remember them. Good scientists are always thinking about whether they are working safely.

Activity – safety first

Here are some basic rules for working safely in science. For each rule, discuss why it is important and what could happen if you did not obey it.

1 Always listen to instructions and follow them carefully.

2 Do not play around with science equipment. Always use it carefully and sensibly.

3 Never run around when doing science experiments.

4 If you are given goggles or safety glasses to wear, keep them on until you are told that it is safe to remove them.

5 Keep all substances and equipment away from your mouth.

Look at the picture on the next page. Some of the children are working safely and some are not. With your partner or group, decide which children are acting safely and why. Then find the children who are not acting safely and say what they should do differently.

Now you know how to be a good scientist. You are ready to learn lots of new things. Have fun!

Life processes

Making groups

In science we often need to sort things into groups to help us to study them. When we put things into a group we look for features that they have in common. For instance, when sorting a group of shells, we might look at their shapes. We could put all the cone-shaped shells in one group and all the spiral-shaped ones in another.

■ We can sort seashells into groups by looking at the different shapes

Sorting objects

Look carefully at these pictures.

robot   star   girl   car

ladybird   water   grass   magnet

snake   tree   cloud   owl

stone   dog   book   fish

Discuss with your partner or group what feature you could use to sort these things into two groups. How many different ways of sorting can you think of?

Choose your best idea and then make a list with two columns and write each item in the correct column. Do not write down which feature you used to sort them.

Show your list to the rest of the class and see if they can guess what your sorting feature was. Can you guess what feature other groups used?

Now discuss everyone's ideas with the whole class. Were some ideas easier to guess than others? Can you suggest why these were so easy?

We sort things into groups to make it easier to study them. It is helpful if the sorting feature can be easily seen and understood by everybody.

➲ Alive or not?

The study of living things is called **biology**. We could sort the objects in the pictures according to whether they are alive or not.

Activity – identifying living things

In your book make a table with two headings: 'Alive' and 'Not alive'.

Look carefully at the pictures again. Write each object into the correct column to show whether it is alive or not.

Now discuss with your partner or group how you can tell which things are alive and which are not. What features do we find in all living things?

Share your ideas with the rest of the class. Did you all use the same features to sort the objects?

All the animals and plants in the pictures are alive, but all living things will die sometime. Discuss with your partner how you can tell the difference between something that is alive now and something that was once alive. How can you tell the difference between something that was once alive and something that has never been alive?

Did you know?

Scientists have a special word for things that are alive. They call them **organisms**.

➲ Life processes

There are millions of different kinds of living things on Earth. Some are enormous, such as an elephant or a giant redwood tree. Some are very small, such as insects and **microscopic** water plants. Sometimes it is easy to tell if something is alive but sometimes it is quite hard.

All living things carry out certain important activities to keep them alive and healthy. We call these **life processes** and scientists will look for life processes to tell if something is alive. Four important life processes are **movement**, **growth**, **nutrition** and **reproduction**.

Movement

Only living things can move by themselves. The car in the pictures that you were sorting earlier can move but only when we put in some fuel and start the engine. It cannot move all by itself.

We can often see animals moving. Different animals move in different ways. How many different ways can you think of?

Activity – animal movement

Find out more about how animals move. Make a list of all the different ways and give some examples of animals that move in each way. Remember to think about animals moving in water and in the air as well as on land.

Make a poster or wall display about animal movement.

Go further

Plants can move too, even though we do not see it happening. A sunflower can turn its flower round so it is always facing the Sun. Plants can move their leaves so that they catch as much sunlight as possible.

■ The flowers of these sunflowers have all turned to face the Sun

Did you know?

Some plants move parts of themselves at different times of day or to protect themselves. Daisies close their flowers at night. The Sensitive Plant, properly known as *Mimosa pudica*, is able to fold its leaves up very quickly if it is touched. The sudden movement and apparent disappearance of the leaves startles grazing animals and helps to protect the plant from being eaten.

Nutrition

All living things need energy. They also need the materials to build and to repair their bodies. These important things come from food. Nutrition is the process of obtaining food. Green plants make their own food (see Chapter 2). Animals need to find the right foods to eat to make sure that they take in all the important raw materials that they need. Some animals eat just plants, some eat other animals and some, like most humans, eat both. You will learn more about how to choose the right foods to eat in Chapter 3.

Growth

All living things grow. When you were a baby you were very small. Now you are larger and you will continue to grow for a few more years. A young oak tree growing from an acorn is tiny but it will grow into a giant tree many metres high.

Reproduction

Reproduction is the process that makes more of a certain **species** (type of living thing). Animals lay eggs or give birth to babies. Plants make seeds that grow into new plants. All living things need to reproduce to make sure that their species does not become **extinct**.

■ A tiny oak seedling will grow to make a huge new oak tree

Did you know?

Giant pandas live in China. They only eat bamboo and they are in danger of extinction because there are not enough bamboo forests left to support them. They are also very slow at reproducing. Each female panda can only have a baby once every two years at most and may have fewer than five babies in her whole life.

Rabbits are very fast breeders. A female rabbit can begin to have babies at the age of 6 months. She may have 8 to 12 babies in each litter and could have more than 10 litters in one year. That's a lot of rabbits!

Bacteria are even speedier at reproducing. Bacteria are made up from just one **cell**. They do not have babies but reproduce by splitting themselves in two. One bacterial cell can become two after just 20 minutes. If all the cells survived, one bacterial cell could become as many as a million in just six hours!

Activity – be a researcher (life processes)

Choose an animal and find out about how it moves, what it eats, how big it becomes and how it reproduces. It will be much more interesting if you choose an animal that we do not see every day.

Put your findings together to make an information poster. Maybe your class could make a display of your research so that other people can learn about your animals too.

Remember to show on your poster where you found the information and make sure that you write it all in your own words.

If you have time you could do the same research about a type of plant.

Go further

➲ More life processes

We have learnt about four of the life processes that we can see taking place in all living things. Here are three others. You will learn more about these later.

Respiration

Respiration is the way that living things release energy from their food. They usually need oxygen from the air to do this. Respiration takes place inside all the cells in the body of every living thing.

Sensitivity

Living things can sense what is going on around them. They then react to what they sense to keep themselves safe, find food and search for a mate. You have five senses: sight, hearing, smell, taste and touch. Living things sense their environment in many different ways.

Excretion

All living things produce waste materials in their bodies. Excretion is getting rid of this waste before it poisons the body.

1 Life processes

There are seven life processes:

Movement

Reproduction

Sensitivity

Nutrition

Excretion

Respiration

Growth

To remember them we often use a memory trick called a mnemonic. Look at the first letters of the seven words. They spell

M R S   N E R G.

Whenever we want to remember all the life processes we can think of Mrs Nerg and she will help us.

Exercise 1.1a

Use the following words to complete the sentences.

climbing extinct flowers food growth larger leaves life processes movement nutrition reproduction running swimming

1 All living things carry out _____ to keep themselves alive and healthy.

2 All living things need _____ for energy and growth. The process of obtaining food is called _____.

3 A young animal or plant is small but it will become _____ as it becomes older. This process is called _____.

4 Living things need to carry out the process of _____ to make more of their species so that they do not become _____.

5 Animals can change their position by _____, _____ and _____. This is called _____.

6 Plants may move their _____ or _____ to face the Sun.

Exercise 1.1b: extension

1 What word do scientists use to mean 'living things'?

2 Which life process gets rid of waste materials from the body?

3 What is the name of the process that living things use to get energy from food?

4 What is meant by the term 'sensitivity'?

5 Some snakes can sense their prey at night even when it is too dark to see. Find out how they can do this.

Did you know?

Sometimes it is very hard to tell whether an animal is alive or not. Animals that hibernate often carry out life processes so slowly that it is almost impossible to believe that they are still alive.

In the Arctic, animals have to survive very, very cold conditions while they are hibernating. Hibernating mammals, such as polar bears, are insulated by thick fur but other animals do not have fur so how do they survive?

The Alaskan wood frog actually freezes solid during the winter and then thaws out again in the spring. If you froze solid, you would die. The Alaskan wood frog manages to survive by filling its cells up with sugar.

Green plants

Plants are very important. If there were no plants there would be no life on Earth. There would be no food and no oxygen in the air. In this chapter you will find out a bit more about these wonderful living things.

⬦ Parts of a plant

There are many different types of plant. Most of them have certain important parts and each part has an important job to do.

Here is a diagram of a flowering plant.

What is the role of the flower?

The flower is needed for reproduction.

The flower makes the seeds that will grow into new plants.

What is the role of the leaves?

The leaves are the places where the plant makes its food (see 'Plant food' later in this chapter).

They are usually green because they contain a special green **pigment** (colour) called **chlorophyll**. This catches light energy from the Sun.

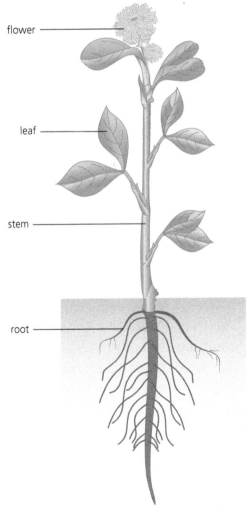

■ The main parts of a flowering plant

What is the role of the stem?

The stem has three important jobs.

- It holds the flower in a good position for **pollination**.
- It holds the leaves in a good position to absorb light from the Sun.
- It carries water and **mineral salts** around the plant.

What is the role of the roots?

The roots hold the plant firmly in the ground. They also take in water and mineral salts from the soil.

Activity – looking at real plants

If possible, take a walk around your school and find as many plants as you can. If you cannot do this, your teacher will make sure that there are some plants in the classroom for you to look at.

On each plant identify the stem and leaves. What differences do you notice between the leaves and stems on the different plants? See if you can suggest any reasons why these differences occur.

The stems of large plants become very strong and woody. How do you think this helps the plant? What name do we give to the big stem of a tree?

Have a look for any flowers on the plants. Sometimes flowers are easy to spot and sometimes they are very small and green so you may have to look carefully for them.

You may not be able to see the roots of most of the plants. Why is this? Look carefully at your plants because you may be able to find some roots that you can see. Plants usually have lots of little roots rather than one big one. See if you can suggest two reasons why lots of little roots might be better.

Exercise 2.1

Use the information above to help you say which part of the plant (flower, leaf, stem, root) carries out each of these jobs:

1 makes food

2 carries water round the plant

3 makes seeds for reproduction

4 takes in water and minerals from the soil

5 holds the plant in the ground

6 holds the leaves and flower in the best positions.

⮕ Growth

Growth is one of the life processes you learnt about in the last chapter. Can you remember the other life processes?

Plants need certain things to grow. If they do not have the right conditions they may become weak and may even die. What conditions do you think plants need to stay healthy?

Poor plants!

For this activity you will need four pots of **seedlings** such as mustard and cress.

Control

One pot is going to be your happy seedlings. Put this pot on a sunny windowsill and remember to keep it well watered. This is what we usually do when we grow plants and we know that it works well. We call this the **control**. We can compare what happens in the other pots with this one to see what has happened.

Working Scientifically

Dark

Put one of your pots of seedlings in a dark place such as a cupboard. Remember to keep it watered but keep the cupboard door closed at other times. What do you think will happen to these seedlings in the next few days?

Dry

Put the third pot of seedlings beside the one on the windowsill but do not water it. What do you think will happen to these seedlings in the next few days?

Cold

Put the last pot of seedlings in a cold place such as a fridge. Remember to check whether the seedlings need watering but put the pot back in the fridge afterwards. What do you think will happen to these seedlings in the next few days? (Note: the inside of the fridge will also be dark so you will need to compare these seedlings with the ones in the dark cupboard.)

Leave your seedlings for a few days to see what happens. You could record the changes by taking photographs each day. After about a week you should be able to see some differences between the seedlings in the different conditions. What does this tell you about the conditions the little plants need to grow well?

What do you think would happen to your seedlings if you started treating them well? You could put all the pots on the sunny windowsill and water them as you did with the control pot to see if they recover.

In the activity above you saw that the seedlings could not grow well if they did not have the right conditions. Plants need water and light to grow well. They also need to be warm enough. What do you think would have happened if your seedlings were put in a very hot place? Maybe you could try it out to test your prediction.

There are many different types of plants living in all sorts of different habitats. All of them need light, water, air and the right temperature to grow well, but in other ways their needs may differ. Some plants grow in hot deserts where there is very little water. They often have thick rubbery leaves and special stems where they can store water to keep themselves alive. Plants like grasses can live very close to each other, but other plants, like big trees, need more space. Some plants will grow in almost any type of soil but others are more fussy, perhaps only growing well in very rich soil where there are lots of nutrients.

➔ Plant food

Plants can make their own food in their leaves. To do this they need water and a gas called **carbon dioxide**. They also need energy.

The water comes from the soil. Can you remember which part of the plant takes in the water and how it reaches the leaves?

The carbon dioxide comes from the air. It gets into the leaves through microscopic holes called stomata in the underside of the leaf.

The energy comes from the Sun in the form of light. The green pigment in the leaves (chlorophyll) takes in the sunlight and the plant uses it to make sugars from the water and carbon dioxide. The plant stores the sugars as a supply of food that it can use to provide energy and materials for growth. This food-making process is called **photosynthesis**.

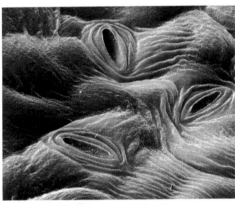

■ Tiny holes (stomata) allow gases to move in and out of the leaf

Did you know?

When plants carry out photosynthesis they make oxygen as well as sugars. They need some of this oxygen for respiration (see Chapter 1) but most of it goes into the air.

Other living things take in the oxygen from the air for respiration and give out carbon dioxide. This helps the plants because it provides more carbon dioxide gas for the plants to use to make food.

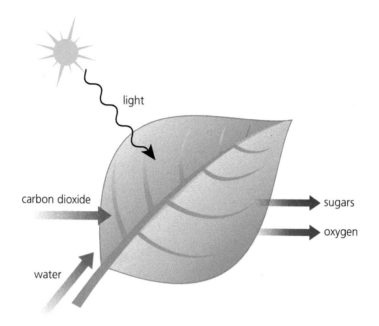

light

carbon dioxide

water

sugars

oxygen

■ Plants can make food in their leaves using carbon dioxide, water and light energy

Did you know?

In the rainforests of the Amazon the trees are so close together that very little light reaches the ground. Plants that grow on the forest floor usually have enormous leaves to help them trap as much sunlight as possible.

Moving water and food around the plant

The roots of a plant are very important. We know that they anchor the plant safely into the ground. They also take in water and certain important nutrients from the soil. These nutrients are called mineral salts and the plant only needs tiny amounts of them to stay healthy. We also need minerals in our diet to stay healthy. You will learn more about these in Chapter 3.

Take a look at this picture of the roots of some plants.

You can see how the roots branch out into the soil. The main root branches out into lots of little roots. This means that the plant can find water and mineral salts from a much greater area of soil.

All parts of the plant need water and mineral salts from the roots. All parts of the plant also need food from the leaves. Plants need a transport system to carry these things around from one place to another. They have tubes passing through the stem from the roots to the leaves and flowers and back again. A watery liquid called sap moves around through these tubes carrying important substances to all parts of the plant.

Did you know?

Some plants grow in places where there are very few mineral salts in the soil. They still need mineral salts so they get them by catching insects. Examples of these plants are Venus flytraps and pitcher plants.

■ A pitcher plant

Moving substances around the plant

You will need:

- a stick of celery
- a white flower on a stem about the same length as your celery stick
- a container such as a beaker or jar
- food colouring
- a knife and a board to cut on
- water.

Put some water into the beaker or jar and add some food colouring to make a strongly coloured solution. Do not make the water too deep. Only about a quarter of the length of your celery or flower stem should be in the water.

Make sure that you know which way up the celery stick should be. Trim off the bases of the stems of the celery and the flower and quickly put them into the water. Ask an adult to help you with this.

Leave the container in a safe place overnight and take a look at it the next day. What has happened? Record your observations using words and labelled diagrams.

Take the celery stick out of the water and carefully cut a slice across the stem. Look at it carefully. You should be able to see coloured dots that show where the special tubes pass through the stem. Draw a labelled diagram to show what you can see.

If you have a spare white flower with a fairly thick stem, your teacher may be able to split the stem into two for part of its length. You can then try putting one half of the stem into one colour of water and the other half into a different colour. Predict what you think might happen. Leave it overnight and see if you were right.

Exercise 2.2a

1 What is the name of the process used by plants to make their own food?

2 a What is the name of the green pigment in leaves?

b What does this green pigment do?

3 Plants need water to make food. How does the water reach the leaves?

4 Plants also need carbon dioxide from the air to make food. How does this gas get into the leaves?

5 Plants need tiny amounts of special substances from the soil to stay healthy. What do we call these substances?

Exercise 2.2b

Use the following words to help you to fill in the gaps in these sentences.

**chlorophyll   holes   mineral salts   photosynthesis   roots
soil   stem   sunlight**

1 Plants carry out _____ to make their own food.

2 Leaves look green because they contain a pigment called
_____. This pigment absorbs _____.

3 Water is taken in from the _____ by the _____
and moves up through tubes in the _____ to the leaves.

4 Carbon dioxide enters the leaf through tiny _____ in the
under surface of the leaf.

5 To stay healthy, plants need to take in tiny amounts of
_____ from the soil.

Exercise 2.2c: extension

1 Plants take in carbon dioxide when they make their food. Describe one
way in which this gas gets into the air.

2 Which gas do plants give out into the air when making food?

3 What would happen to life on Earth if the Sun stopped shining? Explain
your answer.

⊙ Reproduction in flowering plants

We learnt in Chapter 1 that one of the life processes carried out by all living
things is reproduction. The part of a plant that carries out this process is the
flower. On the next page is a diagram of the inside of a flower.

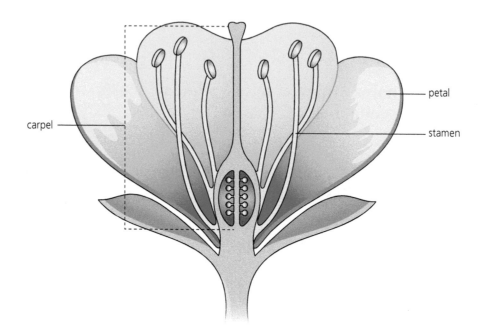

carpel

petal

stamen

Each part of the flower has a special job to do. The male part of the flower is called the **stamen**. The part at the top of the stamen makes **pollen** grains. The female part of the flower is called the **carpel**. At the base of the carpel are some tiny eggs. These will become seeds, but first they need to join up with the pollen cells, usually from another flower of the same type. For this to happen, the flower needs to use a messenger to carry the pollen from one flower to another. For many plants the best messengers are insects.

The plant attracts the insects with brightly coloured petals and a sweet liquid called nectar. This is kept at the bottom of the flower. The insect flies to the flower and crawls down inside to reach the nectar. Pollen from the stamens gets stuck on the insect's body. When the insect has finished eating the nectar it flies off to another flower to find some more, carrying the pollen with it.

The top of the carpel has a sticky surface on it. When the insect reaches the next flower some of the pollen will brush off its body onto the sticky surface. This transfer of pollen from one flower to another is called pollination. The pollen grain then burrows down through the carpel until it reaches the egg at the bottom, and the pollen and egg join together to make a seed.

The seed takes time to grow properly. While this is happening the petals and stamens fall off the flower and the carpel changes to make a seed case or fruit.

Many plants, such as grasses, do not use insects to carry the pollen. They are pollinated by the wind. These plants have small green flowers, often hanging down to blow in the wind. They have to make lots and lots of pollen because so much of it blows away without meeting another flower.

■ A fruit such as an apple contains seeds that could grow into new plants

Exercise 2.3a

1 Which part of a flower makes pollen?

2 Which part of a flower contains eggs?

3 What do we mean by the word 'pollination'?

4 Describe how some flowers attract insects.

5 Describe how an insect pollinates a flower.

6 Some flowers are pollinated by the wind. Describe how you might recognise a wind-pollinated flower.

7 Describe what happens in the flower after it has been pollinated.

Exercise 2.3b

1 Match each flower part to its job.

stamen	attracts insects
petal	contains eggs
carpel	makes pollen

2 Copy these sentences and fill in the gaps using the following words.

egg   insect   pollen   pollination   seeds   wind

a To make a seed, a grain of _____ must join with an _____.

b Pollen can be moved from one flower to another by an _____ or by the _____. This process is called _____.

c After pollination the eggs turn into _____.

Seed dispersal

You saw from your experiments earlier in this chapter that plants need water and light to grow. When a plant makes seeds it needs to make sure that the young plants grow somewhere where they will get enough water and light. If the seeds just fell to the ground the plants would be crowded together and there might not be enough water to go around. They would also be growing in the shadow of the parent plant. The seeds need to travel away to a new place to grow successfully. This is called **seed dispersal**.

Every plant sends its seeds out into the world using a different method. After the egg and the pollen have joined together to form a seed, the carpel changes shape to make a seed case or fruit that will help the seeds to move.

Using animals to disperse seeds

Some seeds are held in sweet, juicy fruits such as blackberries or blackcurrants. Birds and other animals enjoy eating the fruits and the seeds are passed out in their droppings a little while later.

Other seeds, such as nuts and acorns, contain food inside them. Animals will nibble these and destroy them. However, some animals, such as squirrels, may store nuts and acorns for winter food by burying them in the ground. Some of these will be left uneaten in the ground and may grow into new plants.

■ Some of the nuts buried by the squirrel will grow into new plants

Another way to move seeds from one place to another is to hitch a ride. The seeds of some plants, such as burdock, have tiny hooks on them. When an animal brushes past them they get caught in the animal's fur and carried away until they get scratched off later.

Using wind to disperse seeds

Some seeds, for example orchids, are very small and light. When the seed case opens, these can be carried away on the wind. Other seeds, such as those from the dandelion and thistle, are too heavy to be blown around like this, so they use their own parachutes instead. These hold the seed up in the air so it drifts in the wind. Sycamore seeds have little blades on them that make the seeds spin like a helicopter. Some plants, such as poppies, have seed cases like pepper pots with small holes at the top. When the wind blows the seeds are shaken out.

■ The parachutes on dandelion seeds help them to be blown away by the wind

Using water to disperse seeds

Some plants that grow near water have waterproof seed cases that float. These can drift for long distances down rivers and across the sea and be washed up on a distant shore. Coconuts, which are some of the world's largest seeds, are dispersed this way.

■ The coconut seed can float for miles across the sea

Seed cases that explode

Seeds that grow in pods, such as gorse and witch hazel, are often dispersed by a mini explosion. The seedpod dries and becomes so tight that it bursts open, flinging the seeds away into the air. If you walk on a gorse heath on a sunny day in summer you can often hear the seed pods popping open as you pass.

■ The witch hazel pod explodes to disperse the seeds

Investigating seed dispersal

1 Looking at fruits

Make a collection of different fruits. You could include some, such as tomatoes and peppers, that we do not usually think of as fruits.

Ask an adult to help you to cut each one in half carefully so that you can see the seeds inside.

Working Scientifically

Look carefully at the seeds and see how many different things you can observe about them, for example how many there are, their shape, size and colour, and how they are arranged in the fruit.

Make a careful drawing of each one and label it neatly.

2 Sorting seeds

Remember that the seeds of many plants are poisonous. You should never put seeds or berries in your mouth without checking with an adult first. (Birds and other animals can safely eat fruits that are poisonous to us.) Wash your hands immediately after doing this activity.

Collect as many different seeds in their seed cases or fruits as you can.

Sort them into groups according to how they would be dispersed.

Discuss with your partner or group how each one is specially shaped to help the seeds to move to a new place.

Record your observations using drawings or photographs.

The sweetie seed challenge

You will need:

- a sweet to act as the seed
- a range of different materials such as paper, card, thread
- sticky tape and glue
- a stop watch or timer
- a tape measure or metre rule.

Your challenge is to make a seed case that will keep your 'sweetie seed' in the air for as long as possible. Think about the seed cases you have looked at. You might want to model your seed case on one of these or you might decide to try something completely different.

When you have made your seed case, ask an adult to help you to try it out safely by dropping it from a height. How long does your seed case stay in the air? How far does it travel?

Compare your seed case with those made by the rest of your class. Which one stays in the air longest? Which one travels furthest? Can you explain why these ones are so successful?

Exercise 2.4a

1 What does the term 'seed dispersal' mean?

2 Explain why plants need to disperse their seeds.

3 Describe how the seed cases of a dandelion help it to disperse its seeds.

4 Why do some plants make juicy fruits round their seeds?

5 Explain how a squirrel might help an oak tree to disperse its seeds.

Exercise 2.4b: extension

1 Look at these pictures of seed cases. For each one, suggest how they help the plant to disperse its seeds. Explain your answers clearly.

■ Burdock

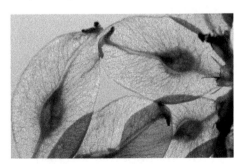
■ Elm

■ Pomegranate

■ Horse chestnut

2 Find out how a peanut plant disperses its seeds.

New plants

When the seeds reach their new home they will wait for the right time to start growing into new plants. Each seed contains a tiny baby plant and a store of food. The little plant needs this food to give it energy to start growing and to keep it growing until it is able to start making food for itself.

We call the first stage of growth **germination**. The first thing to grow is the root. It pushes out of the seed and burrows down into the soil. Here the root can start to take in water and mineral salts from the soil. The root continues to grow and helps to fix the plant in the ground.

Germinating seeds

You will need:

- a clean glass jar
- a roll of blotting paper or kitchen paper
- two or three broad bean seeds
- water.

Soak the seeds in water for 24 hours.

Roll up the blotting paper or kitchen paper and slide the roll into the jar.

Carefully wedge the seeds between the paper and the glass, about half way down the jar, so you can see them clearly.

Pour some water into the bottom of the jar, just enough to wet the paper but not deep enough to touch the seeds. Put the jar somewhere safe.

Now wait and observe your seeds over the next few days.

Measure the length of the root and the shoot each day after they have started to grow. Record your results in a graph to show how much your plant has grown each day.

Don't forget to keep your plants watered.

The WOW factor

When seeds are dispersed they need to wait until the conditions are right before they germinate. Farmers and gardeners need to provide the right conditions for their seeds so that they can grow their crops and flowers. What do you think seeds might need to make them germinate? Discuss this with your partner or group. Remember to explain your ideas clearly using your scientific knowledge.

WOW! Growing seeds

Now set up an experiment to test your ideas about what conditions seeds need to germinate.

You will need:

- seeds – mustard seeds are good because they germinate quickly
- five beakers
- blotting paper or kitchen paper
- silica gel crystals
- cling film
- water that has been boiled and cooled
- oil.

Start by setting up your control experiment.

Put a layer of blotting paper or kitchen paper in the bottom of a beaker and make it wet. Pour out any extra water.

Count out some seeds. If you are using mustard seeds about 20–25 will be enough. Your teacher will help you to decide how many you should use. It is a good idea to use the same number of seeds in each of your experiments. Can you suggest why?

Sprinkle the seeds onto the wet paper in the beaker, cover the beaker loosely with cling film and put it in a warm, light place.

These seeds are warm and they have air and water and light. We expect these to grow well because this is how we grow mustard seeds for salads. This is the control for this experiment. We can compare the other experiments with this one.

Now you can set up your test experiments.

Set up another beaker, exactly the same as the control.

Place it in the fridge where it is cold.

Set up a third beaker, exactly the same as the control.

Place it in a cupboard where it is dark.

To see if the seeds need water, put a layer of silica gel crystals in a fourth beaker then a layer of dry paper on top. The silica gel will take any moisture out of the air so the seeds are really dry.

■ The control seeds have plenty of water, air and warmth

Put the seeds on the paper and then cover tightly with cling film to stop any water getting into the beaker.

Put this beaker with the control experiment.

What about air? You know that living things need oxygen from the air to get the energy from their food. Do you think that the seeds also need oxygen to germinate?

There is air dissolved in water. We know this because fish and other water animals can take in air from the water through their gills. If we boil the water, all the air comes out so the boiled water that your teacher gives you has no air dissolved in it.

Put the seeds in a fifth beaker and cover them with the boiled water.

Pour a thin layer of oil on top of the water. The oil makes a seal between the water and the air to stop any oxygen from the air getting into the water.

Put this beaker with the control experiment.

If your group has another idea about what the seeds will need, ask your teacher if you can set up another beaker to test your idea.

Check your seeds every day. Remember that you are looking for little roots coming from the seeds. After two or three days, write down how many of the seeds in each beaker have germinated.

Look at your results.

Did the cold seeds germinate as well as the control? Do seeds need warmth?

Did the dark seeds germinate as well as the control? Do seeds need light?

Did the dry seeds germinate as well as the control? Do seeds need water?

Did the seeds in the water with no air germinate as well as the control? Do seeds need air?

In these experiments you should have found that seeds will wait until they have water, oxygen and warmth before they start to grow. The first letters of these three conditions spell WOW! This is a good way to remember what conditions seeds need to germinate successfully.

Seeds do not need light to germinate. When we plant seeds, we bury them in the soil where it is dark. The seeds would not grow if they needed light. A seed that has been buried will stay moist and is not so likely to be eaten, so it is helpful that they do not need light to make their first root.

When the seeds have germinated, the next part to grow is the shoot. This is the part of the plant that will become the leaves and stem. This pushes up out of the soil towards the sunlight. As the shoot grows it quickly sprouts leaves, which open wide to trap the sunlight using the green chlorophyll. The little plant can now make its own food. It will grow more leaves and the stem will get taller and stronger. Soon this new plant will grow flowers and start to make seeds of its own.

We call this story of the life of a living thing a **life cycle**. You will learn more about life cycles in Year 5. Here is a picture of the life cycle of a plant.

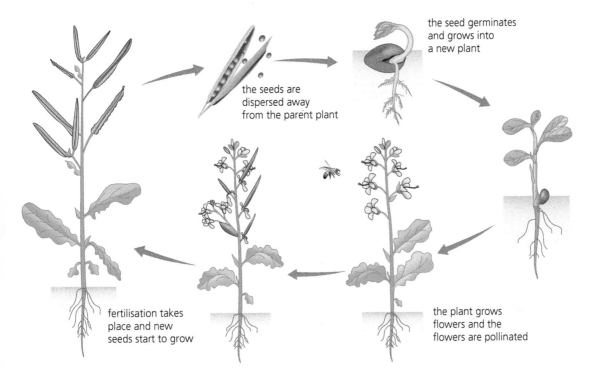

the seed germinates
and grows into
a new plant

the seeds are
dispersed away
from the parent plant

fertilisation takes
place and new
seeds start to grow

the plant grows
flowers and the
flowers are pollinated

 Exercise 2.5

1 What does the word 'germination' mean?

2 Explain why seeds contain a store of food.

3 What three conditions must be right for a seed to germinate?

4 Suggest why the root grows first, before the shoot. Try to give at least
two reasons.

A healthy diet

Our bodies

The human body is like an incredibly clever machine. It is made up of millions of tiny living parts called cells. Each cell is specially made to do a particular job. The 'human body machine' mends itself when it is damaged. Its control centre, the **brain**, keeps it all working properly. It is truly amazing.

You need to look after your own 'human body machine'. To help you to do this, you need to learn more about what it contains and how it works.

What is inside?

The cells that make up our bodies are grouped together to make working parts, called **organs**. Some of the important organs that we will be learning about are shown on this diagram.

Each of these organs has a special job to do.

The brain is the control centre for our bodies.

The **heart** pumps the **blood** around the body. Blood carries important substances such as oxygen and dissolved food to our cells.

The **lungs** take in oxygen from the air and get rid of carbon dioxide from our bodies.

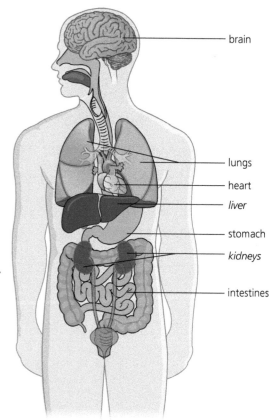

- brain
- lungs
- heart
- *liver*
- stomach
- *kidneys*
- intestines

■ Organs inside the human body

The **stomach** and intestines **digest** our food so that it can be used in our bodies.

The **intestines** also take the digested food into the blood so that it can be carried to the cells.

 Go further

The liver has many jobs, including clearing some poisonous substances, such as alcohol, out of our blood.

The kidneys clear poisons and excess water out of the body.

There are many other organs in our bodies and you will learn more about them later.

All these important organs are safely held inside a layer of **flexible**, waterproof skin. The whole body is supported and protected by a strong **skeleton** (see Chapter 4).

 Did you know?

The brain sends messages to the rest of the body by sending tiny electrical signals down a 'wiring system' of nerves.

 Did you know?

An adult's intestines are about 10 metres long. They are tightly looped up to fit into the body.

Just a minute

Find out how many times your heart beats in one minute. You will need to find a place on your neck or wrist where you can feel the regular pumping of blood through an artery. This is called a pulse point.

When you have found it, count the beats whilst your partner times you for half a minute (30 seconds).

Working Scientifically

Work out how many beats there would be in one whole minute by multiplying your counted beats by 2.

Compare your pulse rate with those of the rest of your class. Did you all count the same number of beats? If not, see if you can suggest any reasons why different people might have counted different pulse rates.

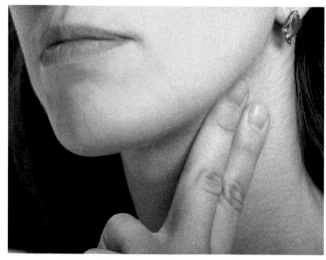
■ You can feel your pulse on your neck

Exercise 3.1

Use the following words to help you to fill in the gaps in these sentences.

carbon dioxide   cells   food   heart   organs   oxygen   stomach

1 The working parts inside the body are called _____ .

2 These are made up of millions of _____ .

3 The lungs bring _____ into the body and send _____ out of the body.

4 The _____ is where our food is digested.

5 The _____ pumps blood around the body to carry _____ and digested _____ to the cells.

➜ Fuel for our bodies

Like all machines, our bodies need fuel to keep them working properly. This fuel comes from the food we eat. Unlike plants, we and all other animals cannot make our own food in our bodies. We need to eat foods that provide us with all the energy we need to be active.

We also need the materials the body uses to build new cells for growth and repair. Other substances are needed to keep the body healthy.

Our diet

Our **diet** is the range of food and drink that we take into our bodies.

We need to choose the right foods to eat. Foods contain special substances called **nutrients**. Different foods contain different nutrients. We should choose a range of different foods to make sure that we are eating all the important nutrients that our bodies need.

Nutrients for energy

To give us the energy we need, we should eat foods containing two types of nutrients: **carbohydrates** and **fats**.

Carbohydrates

There are two kinds of carbohydrate: starch and sugars. Starchy foods are those such as bread, rice, pasta, potatoes and cereals. These foods provide a good source of energy to see us through the day.

■ Starchy foods are a good source of energy

Sugars are found in sweets, cakes, ice cream and fruit. They release their energy quickly. We should try not to eat too many sugary foods because they cause tooth decay.

Fats

Fats are found in meat, fish, dairy products (milk, butter, cheese and yoghurt) and also in foods such as burgers, chips and crisps.

It is important to have some fat in our diet. We store fat under our skin and it helps us to stay warm. It also stores energy in the body.

■ Sweet foods give us energy but are bad for our teeth

Too much fat can make us **obese** (overweight) and can also cause heart disease, so we must limit the amount of fat that we eat.

Nutrients for growth and repair

The body needs materials to build new cells. These materials come from nutrients called **proteins**.

Proteins are found in meat, fish and eggs. Some people, such as vegetarians, do not eat these foods. We can also get proteins from dairy products and from nuts and pulses (beans and lentils).

Nutrients for health

Vitamins and mineral salts

We need these in our diet to help keep our bodies healthy and working properly. To get enough of these, we need to eat a wide variety of different foods, especially dairy products, fruit and vegetables.

An example of a mineral salt is **calcium**. We need calcium in our diet to build healthy bones. Calcium is found in dairy products such as milk, yoghurt and cheese. If people do not have enough calcium in their diet, they might suffer from a disease called rickets. Their bones do not harden properly and become bent.

■ Fat is needed in the diet but too much is bad for our health

■ Foods containing protein are needed for growth and repair

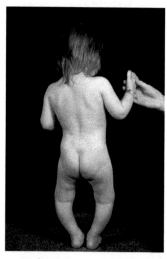
■ Rickets is caused by lack of calcium in the diet

An example of a **vitamin** is vitamin C. Vitamin C is found in fruit and vegetables, especially citrus fruits such as oranges and lemons. Vitamin C helps us to fight off disease. People who do not get enough vitamin C may suffer from scurvy. You can read more about this at the end of this chapter.

■ Citrus fruits provide vitamin C

Water

Water is an important part of our diet. We need plenty of water to keep our bodies working well. Water is found in many foods, especially juicy fruits, but we need to drink as well. Many drinks contain a lot of sugar so plain water is best.

Fibre

Fibre is found in fruits and vegetables. We can also eat it in brown bread and cereals. Fibre is really important to keep the food moving through the intestines properly.

◑ A balanced diet

We have learnt about five different nutrients that we need: carbohydrates, fats, proteins, vitamins and mineral salts. We also need water and fibre. If we eat lots of different foods in the right amounts to give us all these nutrients plus water and fibre, we say that we have a **balanced diet**. It is important to make sure that

■ The 'eatwell plate' shows us how much of each type of food we should eat

we eat a balanced diet as much as possible. It will give us the energy we need for activity and the materials we need to grow and remain healthy.

The 'eatwell plate' shows us how much of each type of food we need in our diet. It shows us that we need to eat lots of fruit and vegetables. We also need enough starchy foods to give us the energy we need. We should eat smaller amounts of the foods containing proteins and fats. Foods with lots of sugar and fat should be eaten only occasionally, as treats.

A healthy, balanced diet

1 Keep a food diary for a day. Write down everything you eat and drink all through the day.

Look at your list and compare it to the eatwell plate in the picture. Do you think that your diet was a balanced one? Is there any type of food that you should have eaten more of? Did you eat too much of any type of food?

2 Design a healthy menu for a day. You should include a good breakfast to give you lots of energy for the day, as well as the other meals and snacks. Remember to make sure that your menu will give you a balanced diet, including plenty to drink.

3 Collect food labels. All food packets must have a label showing how much of each nutrient is contained in the food. Collect a few labels from different foods, for example a loaf of bread, a packet of biscuits, a can of baked beans.

Look carefully at the labels and find out how much sugar is contained in 100 grams of each food.

Draw a table to record your results:

Food	Amount of sugar per 100 g, in g

4 Now show your results in a bar chart.

Which food contains the most sugar? Which one contains the least?

➔ The nutritionist

A nutritionist is someone whose job is to advise people on the best foods to eat. They may carry out research projects to find out how diet helps people to stay healthy or to recover after illness. They use scientific evidence to help people to choose the best foods to eat and to know which ones to avoid.

Some nutritionists work in hospitals and advise patients on what to eat. For example, they may suggest foods that are low in fat for someone with heart disease. Nutritionists also help people who are allergic to particular foods. The nutritionist will help them to find the right foods to eat to stay healthy. A woman who is pregnant might need advice from a nutritionist about how to eat the right foods to help her baby to grow well.

Nutritionists are also a very important part of the training programmes of top sportsmen and women. If a sportsperson is training hard they need to eat the right foods to give them enough energy and to build up their strength. The nutritionist will give them advice on which foods will give them the right balance of carbohydrates, proteins, fats, vitamins and minerals and how much water to drink.

 Exercise 3.2a

1 What is a 'balanced diet'?

2 Name two types of nutrient that we can eat to give us energy.

3 Which nutrients give us the materials for growth and repair in our bodies?

4 Give two reasons why it is important to eat some foods containing fat.

5 Give two reasons why we should not eat too much fat.

6 Why should we eat plenty of fruit and vegetables each day?

Exercise 3.2b

Use the following words to help you to fill in the gaps in these sentences.

**carbohydrates   diet   energy   fibre   grow   nutrients   obese
vitamins   warm**

1 We should try to eat a balanced _____ to provide us with
all the _____ that we need.

2 A good source of energy is _____.

3 Fat is needed as a store of _____ and to keep us _____.

4 Too much fat causes heart disease and may make us become
_____.

5 We need protein to help us to _____.

6 Fruit and vegetables give us _____ and _____.

Exercise 3.2c: extension

A balanced diet is one that provides us with the nutrients, water and fibre
we need to stay healthy. Make a big copy of the diagram below and fill in
the boxes on the right to show what might happen to you if you left out
each of the nutrients named. Water has been done for you. You could do a
bit of research to find out more detail.

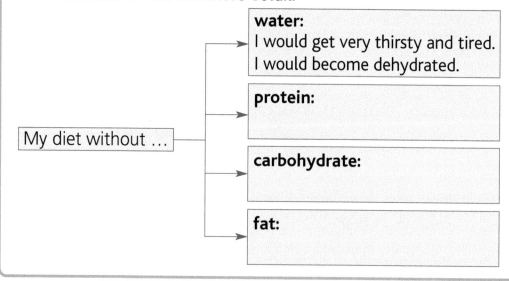

➲ History of science: the story of scurvy

Scurvy is a disease caused by lack of vitamin C in the diet. People knew about the disease nearly 2500 years ago but they did not know what caused it. Scurvy was not properly studied until people started making long sea voyages in the fifteenth century.

On some of these long voyages up to three-quarters of the sailors on the ships died of scurvy. As the sickness usually began after 12 weeks at sea, people thought that it was 'bad air' over the sea that caused scurvy.

When the explorer Vasco da Gama sailed from Portugal to try to find a route to the Spice Islands, in Indonesia, many of his sailors fell ill with scurvy. Sailors with scurvy found that their gums became sore, their teeth began to fall out and their skin became black. Some of da Gama's men died and others were very ill. They were rescued by a ship from Kenya. This ship was carrying a load of oranges and the sick men sucked them hungrily because they could not eat properly. All the men recovered quickly from the disease. The oranges had made them better.

In about 1700, people found a way to preserve oranges so they would last for a long time on the ships. The preserved oranges were a bit like the marmalade you might eat on your toast at breakfast. Sailors could now eat food with vitamin C in it to stop them from getting scurvy. A ship's doctor, called James Lind published a book about how citrus fruit could keep sailors healthy.

Between 1769 and 1771, Captain James Cook sailed from Britain to discover Australia. He took great care over the health of his crew. When he returned to Britain in 1771, not one member of his crew had died from scurvy. This was the first time that a long voyage had ended without any deaths from the disease.

It was not until 1932 that scientists finally proved that it was vitamin C in the diet that prevented scurvy. Refrigerators made it possible to keep fresh foods on board ships so sailors could easily be provided with fresh citrus fruits to keep them healthy. People can still develop scurvy if they do not eat enough foods containing vitamin C, so this is a good reason to include plenty of oranges and other fruit in your diet.

■ Captain Cook took great care to keep his crew healthy

Exercise 3.3

Use the following words to help you to fill in the gaps in these sentences.

air   Australia   black   marmalade   oranges   teeth twelve   vitamin C

1 People can suffer from scurvy if they do not eat enough _____.

2 Sailors often developed scurvy after about _____ weeks at sea. They thought it was caused by bad _____.

3 People with scurvy lose their _____ and their skin becomes _____.

4 Sailors on Vasco da Gama's ship were cured by sucking _____.

5 Some ships began to carry oranges made into _____ to help the sailors stay healthy.

6 The first long sea voyage on which no one died from scurvy was when Captain Cook sailed to discover _____.

Other animals

Other animals often have diets that are very different to ours. Different animals have different diets. For example, a lion eats only meat but a hummingbird eats only nectar from flowers.

■ Lions and hummingbirds have very different diets

If you have any pets you will know that you have to give them the right food to keep them healthy. You cannot feed a dog on hay, and a horse would not be very happy with a can of dog food!

Activity – animal diets

Make a list of animals and find out what they eat.

Can you guess what an animal will eat by looking at it? What clues might you look for? Discuss your ideas with your partner or group.

Write each animal and its diet on a separate piece of paper or card.

Now see if you can sort the animals into groups according to their diets.

We have special words that we use to describe animals with a particular diet. You should know some of them already.

Animals such as cows, horses and pandas, which eat only plants, are known as herbivores.

Animals such as lions, cats and dogs, which eat meat, are called **carnivores**.

We belong to the group of animals that eat meat and plants. We are **omnivores**. Foxes and bears are also omnivores.

Can you guess what an insectivore eats? See if you can find out what a piscivore eats.

Scientists often look at the teeth of an animal to work out what it eats. Different shaped teeth are needed to eat different types of foods. You will learn more about teeth in Year 4.

Skeleton and movement

→ Bones

In Chapter 3 you learnt about some of the important organs in your body. Something else that is in your body is your skeleton. A skeleton is made up from lots of bones. Some are very big, like the bones in your thighs. Some are very small. Inside your ears you have three tiny bones that help you to hear.

Bones are hard and strong. What do you think your body would be like if you did not have a skeleton?

Your skeleton has three important jobs to do:

1 It *supports* your body, keeping it upright and in the right shape.

2 It *protects* the most important organs in your body.

3 It makes a strong structure for the **muscles** to pull on so you can *move* your body.

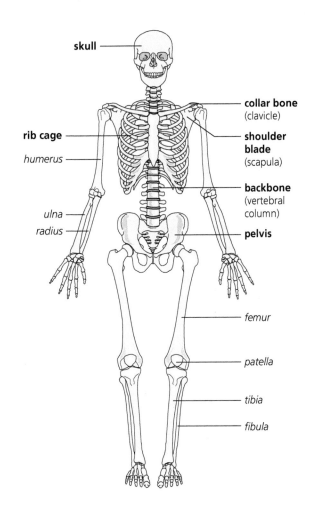

skull

collar bone (clavicle)

rib cage

shoulder blade (scapula)

humerus

backbone (vertebral column)

ulna

pelvis

radius

femur

patella

tibia

fibula

Let's find out a bit more about these important jobs.

Support

Look at the diagram of a human skeleton on the previous page. You can see how all the bones fit together to support the body.

Without a skeleton your body would be all floppy. Your skeleton holds you up so all those important organs inside are in the right place.

Protection

In Chapter 3 you learnt about some of the organs inside your body. The most important of these are the brain, the heart and the lungs. Your skeleton is shaped specially to protect these organs.

Your brain is inside the big bony box called the **skull** to keep it safe. Your heart and lungs are protected by the **rib cage**.

Movement

Your bones move when muscles pull on them. The movement takes place at the places where two or more bones are joined together. These places are called **joints**. You will learn more about joints in Year 6. You have lots of joints. This means that your body can move in many different ways.

Did you know?

Babies are born with about 300 bones. Some of these join together as you grow up, so an adult skeleton has only 206 bones.

Your **backbone (vertebral column)** is made up from 26 small bones called vertebrae. There are joints between each of the vertebrae, which makes your backbone very flexible.

■ Joints allow us to bend our bodies in lots of ways

Activity – no joints

What do you think it might be like if you did not have joints at your elbows and knees?

Try picking up a pencil from the table without bending your elbow.

Try to act out eating a meal without bending your elbows. Can you get the food to your mouth?

Walk around the room normally and then walk around the room without bending your knees. What difference does it make?

Discuss with your partner or group how you would play your favourite sports without elbow and knee joints.

Exercise 4.1a

1 What are the three jobs of the skeleton?

2 Which important organ in found inside the skull?

3 Which part of the skeleton is found around the lungs?

4 Which other organ is found inside this part of the skeleton?

5 What is a joint?

6 Why do we have so many joints?

Exercise 4.1b

Use the following words to help you to fill in the gaps in these sentences.

heart   joints   lungs   muscles   skeleton

1 The bones in my body are joined together to make my _____.

2 My _____ and my _____ are protected by my rib cage.

3 Places where two or more bones come together are called _____.

4 Bones move when they are pulled by _____.

➲ Other animals

Animals that have bony skeletons and a backbone in their bodies are called **vertebrates**. Each vertebrate's skeleton is different but they all have some features in common.

Comparing skeletons

Look at the two skeletons in the pictures below.

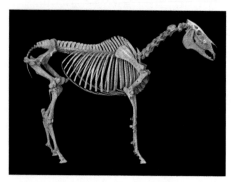

■ The skeleton of a horse

Working with your partner or group, make a list of ways in which these two skeletons are similar.

Now make a second list of ways in which the two skeletons are different.

Next, look at the picture of a human skeleton earlier in this chapter. See how many of the named bones you can find in all three skeletons.

■ The skeleton of an eagle

You have seven small bones (vertebrae) in your neck. A giraffe has the same number!

■ The giraffe's long neck has only seven bones in it

Not all animals have skeletons. In fact, most of the animals in the world do not have skeletons. Think about a worm for example. Its body is soft. It does not have any bones inside. Animals that do not have skeletons are called **invertebrates**.

A small animal does not always need a skeleton to support and protect its body like we do. An earthworm can slither around quite well in the soil without bones because it does not have any legs.

Small animals, such as insects, that have legs to run around on are different. They have a hard case, called an **exoskeleton**, around the outside of their bodies. This makes their legs stronger. It also makes it a little harder for their **predators** to eat them. How does a snail protect its soft body?

■ An earthworm does not have a skeleton

■ Some invertebrates have hard outside cases to protect and support their bodies

Moving around

We have learnt that we can move our bodies when muscles pull on bones. How do animals move when they have no bones?

Animal movement

Start by thinking about the ways in which you can move your body. You might do this in a PE lesson. You could try to move around like different sorts of animal, maybe a kangaroo, a snake and a dog. All these animals are vertebrates. They have bony skeletons inside their bodies moved by muscles.

Now collect some invertebrates from your school grounds or your garden. Remember to be careful not to hurt them. See if you can find an earthworm, a snail and a beetle. You might also be able to find others, such as a centipede, a millipede and a slug.

Discuss with your partner or group what questions you might ask about the ways in which these animals move. Write your questions in your book.

Place each animal in a tray and watch how it moves.

Try to use your observations to explain how each animal is moving its body. See if you can spot the muscles moving the bodies of the animals. If you put a snail or slug on a piece of glass or clear plastic, you will be able to look at it from underneath. If you watch carefully you may see little ripples of movement as the muscles pull its body along.

Use your observations to try to answer the questions you asked. If you cannot answer a question from what you have seen, you could look in books or on the internet to find out more.

When you have finished observing your invertebrates, remember to put them back exactly where you found them and then wash your hands.

Exercise 4.2a

1 Copy this passage, choosing one word from each pair to complete it correctly.

Animals that have internal skeletons are called (vertebrates/invertebrates). These animals can move when muscles (pull/push) on their bones. Animals that do not have internal skeletons are called (vertebrates/invertebrates). Some of these have a (hard/soft) outer shell around their bodies.

2 Draw a table like the one below in your book.

Now look at the following list of animals mentioned in this chapter and write each one in the correct column of the table.

earthworm   beetle   eagle   human   horse   kangaroo   snail   snake

vertebrates	invertebrates

Exercise 4.2b: extension

1 Use books or the internet to find a picture of the skeleton of a dolphin or porpoise. Compare it to the horse skeleton shown in this chapter. Suggest reasons for the differences between the two.

2 Invertebrates are often known as minibeasts. Not all invertebrates are very small. Do some research to find out about some of the larger ones. The largest invertebrates live in the sea. Can you explain why there are no really big invertebrates on land?

Go further

→ Getting bigger

In Chapter 1 you learnt about the life processes that are carried out by all living things. How many of them can you remember? One of these life processes is growth. Most animals are born small and get bigger as they get older. You are much bigger than you were when you were born and you still have a lot of growing to do.

When vertebrate animals grow, their bones get bigger to support their larger bodies. They have soft stretchy skin and it is quite easy to see how the whole body can grow at the same time.

Now think about invertebrate animals, such as crabs, lobsters and spiders, which have hard exoskeletons. Exoskeletons are not stretchy. The animal inside still needs to grow but the skin cannot grow with them so they need a new, bigger one from time to time.

An example of this is a tarantula spider. These are really quite small when they are born, no bigger than a house spider. When they are fully grown they are much, much bigger so they need several changes of skin in their lifetimes. They get bigger and bigger until they are all squashed up inside their skin. The skin splits open and the spider pulls itself out; the old hollow skin is left behind.

■ This tarantula has just shed its skin. You can see the old empty skin beside it

Inside the old skin the spider has grown a new one but it is soft. As soon as the old skin has been shed, the new one begins to harden. The spider needs to hide away when this is happening, as the new soft skin does not provide very good defence from predators. Soon the skin has hardened and the spider can get on with life – and start growing all over again.

Rocks

➜ Inside the Earth

The Earth was formed 4600 million years ago, when bits of dust around the Sun began to join together. If you were able to slice right through the Earth you would see three layers:

- the core in the middle
- surrounded by the mantle
- with a hard outer shell called the **crust.**

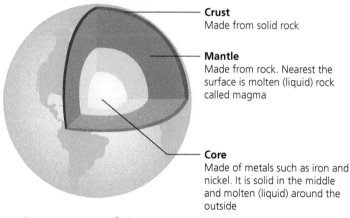

Crust
Made from solid rock

Mantle
Made from rock. Nearest the surface is molten (liquid) rock called magma

Core
Made of metals such as iron and nickel. It is solid in the middle and molten (liquid) around the outside

■ The structure of the Earth

All the rocks that we see around us are part of the crust. They have been made in different ways. Some are very hard; some are softer. Some are made from crystals; some are made from grains.

➔ Rocks made from crystals

Some rocks are made when the liquid rock from the mantle comes to the surface. It cools down and becomes solid rock. The rocks made in this way are made from crystals and they are very hard. If the liquid cools quickly the crystals will be small. If the crystals are big, it means the liquid rock cooled more slowly.

Some crystal rocks are made in a **volcano**. There are many volcanoes in the world and about 1500 of these are active.

A volcano is a mountain that is made when red-hot liquid rock from inside the Earth escapes up through an opening, called a vent.

Each time a volcano erupts, the liquid rock (lava) on the surface of the volcano cools and sets as a solid layer. As the layers build up, the volcano grows larger.

Many volcanoes are under the sea. Some grow so large that they reach up above sea level as new islands. Iceland is a volcanic island. It is still growing with each new eruption.

■ An erupting volcano

Activity – make your own volcanic eruption

By making your own model of a volcano, you can see how a volcano changes its size and shape as lava flows down the sides.

You will need:

- a tray to work in
- play dough in several different colours
- bicarbonate of soda (sodium hydrogen carbonate)
- vinegar.

Put some play dough on a tray and make it into an interesting mountain shape. You will get the best results if you shape your mountain carefully and form ridges and valleys, and make small places where lakes might form and streams might flow.

Make a small hollow at the top of your mountain, no more than 2 cm deep.

Place some bicarbonate of soda into the hollow at the top of your volcano and carefully pour in some vinegar. The mixture will fizz and bubble.

Watch where the bubbles flow, like lava flowing. Now take a small piece of coloured play dough and place it over the areas where the bubbles have flowed, building up that part of your mountain.

Do this several times, using different coloured play dough each time.

Your volcano should now look quite different. You can see how a volcano changes in shape and size as eruptions take place over the years.

→ Pompeii

Pompeii was a Roman town in Italy. It was a good place to live. The surrounding fields were very good for growing crops and people also did well by trading with other places in the Roman world. Many of the people lived in big houses with beautiful paintings on the walls.

When people looked out from Pompeii, they could see the huge cone shape of the mountain called Vesuvius. Vesuvius is a volcano. It had erupted many times. The fields round Pompeii were good for growing crops because the minerals from the volcanic ash made the soil rich and fertile.

One day in 79 A.D. (probably 24 August) started just like any other day. Sometime during the afternoon people noticed a strange cloud rising over Vesuvius. Hot dust and ashes began to fall from the sky. People rushed to the harbour to try to find boats to escape in. Suddenly, there was a great rumble, the mountain seemed to explode and an enormous cloud of hot gas, rock and dust rushed down the volcano and through the streets of Pompeii, burying everyone and everything.

There was so much dust that a layer many metres deep settled over the town. Over time it cooled and settled and turned from dust into rock. Amazingly, although very few people from the town survived and everything was buried under the dust, many of the houses and their contents and even the bodies of the people who died were preserved in the dust.

We know about the story of that terrifying day because a Roman writer, called Pliny the Younger, watched the eruption and wrote about it. His uncle, Pliny the Elder, died helping people to escape in boats from the harbour.

In 1738, some workmen were digging foundations for a new palace for the King of Naples and they discovered some of the remains. Over the years since then, the town has been gradually uncovered. Now it is possible to visit Pompeii and see some of the houses with their painted walls and rich furniture, the streets and temples. We can learn a lot about Roman life from Pompeii.

5 Rocks

Rocks made from grains

Some rocks are formed from **sediments**, which are tiny grains of rock, sand, mud and parts of animals and plants. These are all washed down from the mountains by rivers. They settle in layers at the bottom of the sea or a lake.

As more layers build up, the lower layers of sediment are squeezed by the layers above. This slowly changes them into solid rock. We call these rocks **sedimentary rocks**.

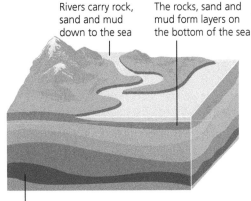

Rivers carry rock, sand and mud down to the sea

The rocks, sand and mud form layers on the bottom of the sea

The layers are squeezed as more material builds up on top of them. They harden to form rock

Activity – make sedimentary rock cakes

Remember to wash your hands before and after this activity.
You will need:

- three different coloured types of biscuits
- three small bowls
- a spoon
- something to crush the biscuits with, e.g. a rolling pin
- melted butter mixed with a little syrup
- paper cake cases.

The biscuits represent different types of sediment. The butter and syrup mixture represents the sticky mud that holds the grains of sediment together in the rock.

Put each biscuit in its own small bowl and then crush it to make 'biscuit sand'. Stir some 'sticky mud' into each bowl so it all sticks together.

Put a layer of one of your 'biscuit sand' mixtures into a paper cake case and press it down firmly with the back of a spoon.

Add a layer of your next 'biscuit sand' mixture and then finish with a final layer of your third 'biscuit sand' mixture.

Press it all down firmly and leave it to cool.

Carefully peel away the paper case to see the different layers in your sedimentary rock cake. You may be able to see them more clearly if you carefully slice down through the cake with a knife. Ask an adult to help you with this.

Fossils

Rocks made from grains (sedimentary rocks) sometimes have **fossils** in them. A fossil is the remains of an animal or plant that died millions of years ago when the rock was being made.

The dead animal or plant fell to the bottom of the sea or a lake. Its body became buried as more sediment fell on top of it. In time the remains of the animal or plant hardened like the rock around it.

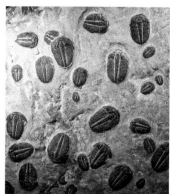

It is usually the hard parts of an animal (bones or shell) that become fossils. Sometimes other softer parts of the body can be seen in the rocks as well. Fossils are interesting because they tell us about animals and plants that lived millions of years ago. They are often very different to the animals and plants on Earth today. Perhaps you could find out about some of the animals found as fossils, such as the trilobites in the photograph. You will learn more about fossils in Year 6.

▬ This rock is full of fossil animals called trilobites

Activity – make a model fossil

Sometimes the actual remains of the animal do not survive but an imprint of their bones or shell in the sediment becomes filled with hard minerals in the exact shape of the animal's remains. You can make a model to show how this happens.

You will need:

● a seashell

● modelling clay

- a strip of card and a paper clip
- plaster of Paris.

Shape some modelling clay in your hand until it is smooth and round. Press it down lightly to flatten it a bit.

Press the textured side of a seashell firmly into the clay and then carefully remove it. It should have left a clear imprint in the clay.

Press a strip of card into the clay to make a wall around the shell shape. Make sure there are no gaps. Fix the overlapping ends of card together with a paper clip.

Your teacher will give you some plaster of Paris mixture. Pour the mixture carefully over the shape left by your shell. The card wall should stop the mixture from escaping.

Leave the mould to set for a few days. The plaster of Paris needs to dry out and harden.

When your model is hard, carefully peel away the card wall and the modelling clay. You can paint your model to make it look like rock and then cover it with clear varnish.

 Go further

⬆ Limestone caves

One common type of sedimentary rock is called limestone. This rock is made up from the shells of tiny sea creatures mixed with sand. It often also contains fossils of larger sea animals.

Limestone is quite a soft rock. Sometimes water under the ground, maybe in an underground stream, gets into cracks and wears the rock away. Some of the rock can dissolve in the water. The cracks get bigger and bigger and over millions of years a cave might form.

Later, rainwater soaks down through the soil and dissolves more of the rock. It seeps into the cave and drips down from the roof. Some of the dissolved rock gets left behind on the roof. Some may get left on the floor of the cave.

As more and more water drips through, a stalactite begins to form on the roof of the cave. A stalagmite grows where the drips fell on the floor. Over millions of years these can grow huge and make beautiful formations in the cave. Have you ever visited a cave full of stalactites and stalagmites?

■ Stalactites and stalagmites form in limestone caves

⮕ Rocks all around us

The Earth's surface is made up from lots of different types of rock. Rocks are very useful materials. We use them for building houses, making roads, carving to make statues. Rocks are also used to make glass. If you search the area around your school or home, you will find rocks in many different places.

■ Rocks are often used to build houses

Looking at rocks

1: rock survey

Look around your home or school and see how many different types of rocks you can find. Remember to look at buildings, walls, pavements and garden statues. You could try to find out where some of the rocks came from. They may be local or they may have come from a long way away.

Look carefully at the rocks. If possible, use a hand lens or magnifying glass to look at them more closely. Are they made from crystals or grains? Can you see any fossils?

Have the rocks been there for a long time or are they new? Can you see any changes that have taken place in the rocks over time? What do you think caused these changes?

Make a record of what you find. Take some photographs and write a short description of each rock and where you found it. Report your observations as a poster or wall display.

2: testing rocks

You will need:

- samples of different types of rock
- a nail
- a bowl of water and a towel
- a hand lens.

Start by looking at the rocks carefully, using a hand lens. Sort them into two groups: rocks made from crystals and rocks made from grains.

Try scratching each rock type with a nail. Does the nail scratch the rock? If so, was it easy to do so or hard? Decide whether you think the rock is hard or soft. Compare the hardness of the rocks in your two groups. Are the rocks in one group harder than the other?

Take two rock samples, hold one in each hand and rub them firmly together. Can you rub bits off the rocks?

Does it make a difference which rocks you use?

Carefully drop each of your rock samples into the water. Look carefully to see if you see any change in the rock when it gets wet. Take them out and put them onto a towel. How easy is it to dry each type?

Discuss your observations with your partner or group. See if you can spot any patterns in your findings. What do these tell you about the two different groups of rock?

Extension

Think about the rocks you spotted in your rock survey. Can you use the observations you made in the testing rocks activity to suggest why each type of rock was used in each place?

Exercise 5.1

1 Join each word on the left to the correct description on the right.

| crust | | the remains of a dead animal or plant trapped in the rock millions of years ago |

| sediment | | a mixture of sand, mud and parts of animals and plants that settles at the bottom of a lake or sea |

| volcano | | a mountain made when liquid rock escapes through a gap in the Earth's crust |

| fossil | | the layer of solid rock on the surface of the Earth |

2 Fill in the gaps in these sentences:

a Rocks may be made up from _____ or _____.

b Rocks made from sand, mud and parts of animals and plants are called _____ rocks.

3 Danny found a fossil of an animal in a rock on the beach. Describe in your own words how the fossil was made.

Soils

➔ What is soil made from?

We know that plants usually grow with their roots in soil. Soil is usually brown. It might be wet and sticky or dry and crumbly.

Soil is found almost everywhere on land. Have you ever looked really carefully at it? Do you think all soils are the same?

■ Soil is needed to grow crops and other plants

Looking at soils

You will need:

- two or three soil samples from different places

- a shallow container or a sheet of newspaper for each soil

- a hand lens or magnifying glass.

■ Different soils may look and feel very different

Working Scientifically

Make sure that you do not mix up your soil samples in this activity. You will need them again later so keep them separate.

Work with a partner or group. Look very carefully at the soil samples. Use your hand lens to look at the different bits and pieces in the soil. Make a list of what you find.

Take a little of each soil and feel it. What happens if you squeeze it? What does it feel like if you rub it between your fingers? What does it smell like? Are there differences and similarities between the soils?

Discuss your observations with your partner or group. Is there anything that you would like to know about the soils? Write down some questions to investigate.

Remember to wash your hands after this activity.

Look at the list you made in the activity. You will have found bits of rock, sand, mud and the remains of plants and animals in your soil samples. How do you think they got there?

The rocky bits

Most of the soil is made from bits of rock. If the pieces are bigger than about 2 mm, we call them **gravel**. There may be smaller grains such as sand. Some are almost too small to see. We call these tiny pieces **silt**.

These rocky bits in the soil have all been made by a process called **weathering**. Rocks are hard and strong but they can be broken down very slowly by the weather. Big rocks may break into smaller ones. These may fall into a stream or river where they tumble around and bump into each other.

■ A fast-moving stream helps to break rocks into soils

The bits of rock get smaller and smaller. In the end they form sand and silt. Some may get left by the side of the river. Some may settle as sediment when the river floods. That is how a big, strong rock on a mountain might end up as soil in your garden.

There are many different types of rock. This means that the rocky part of soil is different in different places. Each soil will be a bit like the rocks in that area. Your soil samples may be different colours, like the ones in the photograph on the first page of this chapter. This tells us that they came from places with different rocks.

Bits of plant and animal

The little bits of once-living things that you found in your soils were probably mostly from dead plants. There will be pieces of dead animal as well but these are often harder to see.

The once-living bits in the soil are very important. They make a material called **humus**. Humus does two very important jobs that help plants to grow well. Humus holds water and adds mineral salts to the soil. Do you remember that you learnt about mineral salts in Chapter 2? Which part of a plant has the job of taking them from the soil?

When the leaves fall off the trees in autumn they provide food for animals such as earthworms and woodlice and for fungi. These living things, known as **decomposers**, break up the dead leaves and other material and mix them into the soil. Gardeners often collect up dead plant material from the garden and kitchen and put it into a compost heap. It will turn into lovely crumbly humus. This can be added to the soil to help plants to grow better.

■ Fungi help to break down once-living things to make humus

Comparing soils

To find out more about soil samples, scientists like to separate out all the different parts so they can compare them. Some soils have lots of bigger pieces such as gravel and sand. Some have lots of smaller, silty pieces.

There are two ways to separate soils to compare them.

1: sieving

You will need:

- a set of soil sieves
- very dry samples of your soils
- access to a balance
- sheets of paper.

■ Soil sieves can be used to separate soil samples

A set of soil sieves has a number of sieves stacked on top of each other. Each sieve has different sized holes. Make sure that they are stacked in order with the biggest holes at the top and the smallest ones at the bottom.

Can you guess what will happen to your soil in the sieves? Why is it important to have them in the right order?

Use a weighing machine (balance) to weigh out equal amounts of each of the soil samples. Your teacher will tell you how much to take. Make sure that you label them so they don't get mixed up.

Can you explain why it is best to have the same amount of each sample?

Put one soil sample into the top sieve and put the lid on. Hold the sieves together and carefully shake them.

Remove the lid and tip the contents of each sieve onto a different sheet of paper. Label them.

Look at the piles on the sheets of paper. Is there the same amount of material on each sheet? If not, does the soil have more large bits or more small bits?

Weigh the material left behind at each level in the sieve. Write the results in a table like this one. If you have more than three sieves in your stack you will need more rows in the table.

Carefully put the sheets of paper to one side.

Do the same for the other samples.

Now compare the results. What differences can you see between the soils?

Sieve layer	Mass of material in the sieve, in g		
	Sample A	Sample B	Sample C
Top layer			
Second layer			
Third layer			
Bottom			

Now plot a bar chart of the results for each soil. How do you think these help us to compare the soils more easily?

Extension

If you are good at drawing bar charts, you might try putting the results of all the samples on the same one. Your teacher will show you how to do this. How might this be even more useful than separate bar charts?

Remember to wash your hands after handling soils.

2: soil soup

If you do not have any soil sieves you can compare soils using this method.

You will need:

- your soil samples
- a jar with a screw top lid for each soil. It is helpful if these are all the same size
- water.

Put some of the soil samples into the jars – a different soil in each jar. The jar should be about a quarter full. Try to make the depth of soil the same in each jar. Why do you think this is important?

Add water to each jar until it is nearly full. Do not fill them right up to the very top.

Screw the lids tightly onto the jars. Label the jars so you know which soil sample is which.

Give each jar a really good shake to mix all the bits of soil with the water. Make sure none has got stuck at the bottom.

Find a safe place to leave your jars for a day or two until the water is clear again. Be very careful not to shake them up again when you take them back to your table to look at them.

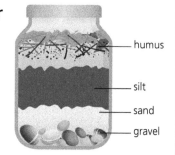

The bits of the soil will gradually sink down to the bottom. The big heavy pieces of gravel will sink quickest, then the sandy bits. The tiny bits of silt take longer to sink. Bits of leaf and stick will float.

Now compare your soils. Look at the thickness of the layers. Are they all the same? Do the different soils have different thicknesses? Is the water coloured? How much is floating on the surface?

Write about your observations. Describe each sample. Write down the similarities and differences between the soils. You could take a photograph of the jam jars or draw a picture to illustrate your work.

Remember to wash your hands after handling soils.

Air and water

When you looked at your soil samples you were looking at the solid bits in them. The spaces between these bits are also important.

All living things need air. The air spaces in the soil allow the plant roots and the animals that live in the soil to get the oxygen they need.

All living things need water too. A healthy soil needs to have some water in it as well. Too much water can be a problem though. It fills up all the air spaces. A healthy soil holds some water but lets the rest drain through.

When you compared your soil samples in the last activity, you will have found that each soil had different amounts of big and small pieces. Look at this diagram.

You can see that a soil that has large pieces of rocky material in it has bigger air spaces than a soil with tiny pieces. Which one do you think would let the water drain through most quickly? Explain your idea carefully.

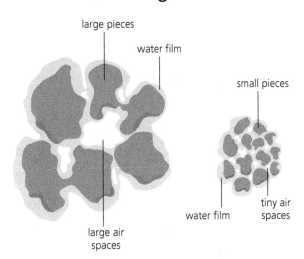

large pieces

water film

small pieces

large air spaces

water film

tiny air spaces

Test the drainage

Working Scientifically

In this experiment you are going to carry out a fair test to find out which of your soil samples lets water drain through fastest.

Think about what you have already learnt about your soil samples. Discuss with your partner or group which sample you expect to drain most quickly and why.

You will need:

- small pieces of disposable cloth (one for each soil sample)
- funnel
- large measuring cylinder

- dry soil samples

- water

- beaker

- stop-clock or -watch.

Place a small piece of cloth into the funnel to cover the hole at the bottom. This will stop your soil falling through and blocking the funnel.

soil sample

cloth to stop soil blocking the funnel

water that has drained through

00:01:36

Sit the funnel in the measuring cylinder so that you can easily measure the volume of water that drains through the soil.

Pour one dry soil sample into the funnel. To make it a fair test, you need to use the same depth of soil for each test, so mark the level you use for the first sample.

Put some water in a beaker. Start the stop-clock and pour the water steadily onto the soil sample in your funnel, so that the water is always kept 'topped up'.

After one minute, stop pouring and quickly measure how much water has passed through the soil into the measuring cylinder.

Record the result carefully in a neat table.

Empty the measuring cylinder. Clean out the funnel, put a fresh piece of cloth in the bottom and repeat the test using another soil sample.

When you have tested all your samples, compare the results. Which soil drained quickest? Was your prediction correct?

This activity shows that soils with big pieces of rocky material in them drain faster than ones with tiny pieces. The large air spaces between the pieces of gravel and sand make it easier for the water to pass through.

Naming soils

When scientists know about the bits and pieces in a soil sample, they can use these to decide what type of soil it is. For example a soil might be described as **sandy soil**, **clay soil** or **loam**.

Here are the properties of these three different types of soil.

Soil type	Size of rocky pieces	Feel when rubbed	Does it stick together when squeezed?	Does it drain quickly?
sandy soil	mostly large: sand and gravel	very gritty	no	yes
clay soil	mostly small: silt	quite smooth	yes	no
loam	an equal mixture of big and small	a little bit gritty	not very well	quite quickly

The best type of soil for growing plants is loam. A good loam soil also contains lots of humus. It holds some water but lets the extra drain away. It has air spaces so the soil animals and plants get enough oxygen. It contains lots of mineral salts to help plants to stay healthy. Gardeners and farmers try hard to make a good loam soil for their plants to grow in.

The activities you have done have given you all the information you need to name your soil samples. Decide whether each soil is sandy, clay or loam.

You have lots of information and results from your investigations into soils. How can you display all this hard work and interesting material so others can find out about soils? Maybe your class could work together to make a bright and informative wall display all about soils. You could try to find out more about the animals that live in the soil as well to add even more interest.

Working Scientifically

Exercise 6.1a

1 Match each soil part to the right description.

gravel	very tiny rocky pieces
sand	remains of dead plants and animals
silt	rocky pieces more than 2 mm across
humus	small rocky pieces, smaller than 2 mm but big enough to see

2 Give two reasons why humus is useful in the soil.

3 How might a gardener make more humus to add to the soil in the garden?

4 Name two important things that are needed in the soil apart from rocky material and humus.

5 Write a short description of how a big rock on a mountain might become a tiny rocky particle in the soil.

6 Why does water drain more quickly through a sandy soil than a clay soil?

7 What name is given to a soil that contains a mixture of different sized pieces and lots of humus?

Exercise 6.1b

1 Write these rocky pieces in order of size, smallest first:

sand   **gravel**   **silt**

2 a What name is given to the material in the soil made from remains of once-living plants and animals?

b Give two ways in which this material helps plants to grow well.

3 Will water drain quicker through soil with big rocky pieces or small ones?

4 Which kind of soil (sandy soil, clay soil or loam) is the best for growing plants?

Exercise 6.1c: extension

Mr Jones is a farmer. He tests the soil in three of his fields. The soil in each field is different. One has very sandy soil. One has a clay soil and one has a loamy soil. He forgot to label his samples so he cannot remember which soil came from which field.

He looks at the fields. In Field 1 there are tall crops growing well. In Field 2 he sees a big puddle where the rain has not drained away. In Field 3 the crops are looking dry. They are not growing well.

Use Farmer Jones's observations to decide which field contains which soil type.

Go further

There is another part of the soil we have not looked at yet. Almost all soil contains lots of living things. There are some living things that are so tiny we can only see them with a microscope. There may be small invertebrates that we do not normally notice and there may also be larger animals, such as worms or moles.

Find life in the soil

You will need:

- a plastic sieve with holes about 0.5 cm in diameter
- a bowl, preferably a white one, that is deep enough for the sieve/colander not to touch the bottom
- a sample of fresh, damp soil
- a lamp
- a soft paintbrush
- a tray or container to put the animals in while you are looking at them
- a hand lens or magnifying glass.

Working Scientifically

Check your soil sample to see if there are any larger animals, such as worms, that could not fit through the holes in your sieve. If there are, carefully take them out and put them back where they live. This experiment could harm them.

Sit the sieve in the bowl and put your soil sample in the sieve.

Then shine the lamp onto the soil. The lamp will need to be quite close to the soil, but not touching it. Leave the lamp on for about one to two hours.

Small animals in the soil like the cool, damp conditions underground. When they feel soil becoming warmer, they move deeper into the soil away from the warmth. This helps to stop them drying out on hot, sunny days.

In this activity, any small animals in the soil will move away from the warmth of the lamp. They will fall through the holes in the sieve and into the bowl below.

Turn off the lamp and carefully remove the sieve.

Use the paintbrush to move the animals gently onto the tray.

Now you can study them closely with a hand lens. How many different types of animal have you collected?

Think about how you can record what you have found. Maybe you could draw neat, accurate pictures of some of the animals. Remember to record how big they are and what colours you can see.

Do the animals have anything in common? If so, can you explain why they are all similar in this way?

Remember to return all animals to their soil and replace the soil sample where you found it.

Wash your hands when you have finished handling the soil.

⊙ Earthworms

Earthworms are very important in the soil because they mix everything together. As they burrow and wriggle in the soil, they make spaces in the soil for water, roots and air.

They pull plant material from the surface deep into the soil. This helps to increase the amount of humus in the soil.

There are lots of different types of earthworm. If you dig up a sample of soil in your garden, you may be surprised at how many different types of worm you can find. You could find a guide on the internet to help you identify them.

We can see how earthworms change the soil by collecting some and placing them in a wormery.

Observe life in a wormery

You will need:

Working Scientifically

- a wormery or large jam jar
- several soil samples
- sand
- leaves
- water
- a few large worms
- black card or cloth (large enough to cover the wormery or jam jar).

Remember to wash your hands after handling the soil, the wormery or the earthworms.

Put a few layers of soil in your wormery with a thin layer of sand in between each layer as shown in the photograph.

Place a few leaves on the surface and pour in enough water to moisten the soil but not to flood it. Put your worms carefully on the surface.

Cover the wormery with black card or a dark cloth and put it in a cool place for a few days. Check now and then to make sure that the soil does not dry out. If it looks a bit dry, add water to the surface, a little at a time, until it looks damp. Make sure that you do not pour in too much or you will drown your worms.

After a few days remove the dark cover. You should be able to see the tunnels in the soil where the worms have been moving up and down the wormery.

Put the cover back on and leave it for a few more days before you look at it again. Over time the layers in the soil will become mixed as the worms burrow up and down.

When you have finished, put your worms back where you found them.

 Did you know?

Some people believe that if you chop an earthworm in half, you will get two earthworms. This is not true. The end of one part of the poor worm may heal and the worm may survive. However, you are much more likely to finish up with two halves of a dead earthworm!

 Exercise 6.2a

Use the words in the box to fill in the gaps in the sentences below.

**air   earthworms   humus   invertebrates   leaves   moles
roots   water**

1 Some small _____ spend their lives in the soil.

2 Larger animals in the soil may include _____ and _____.

3 Earthworms are useful in the soil because their burrows let _____ and _____ into the soil and make space for _____ to grow.

4 Earthworms also pull _____ into their burrows, which increases the amount of _____ in the soil.

Exercise 6.2b: extension

Read the following information about earthworms.

An earthworm spends most of its life in the soil where it is cool and damp. Like many other invertebrates, it needs to keep its skin damp all the time.

It will come to the surface, usually at night, to collect leaves and pull them down into its burrow to eat in safety. It will usually only come out during the daytime if it is raining and its burrow becomes flooded.

As it tunnels through the ground it takes soil into its mouth. The soil passes through its gut and some of the humus in the soil is digested. The rest of the soil is passed out and makes a little squiggly mound called a worm cast. Worms make their worm casts at the top of their burrows on the surface.

Earthworms are eaten by many animals, such as birds, hedgehogs and badgers. Earthworms have little bristles on the sides of their bodies, which they can use to hold themselves into their burrows when they are attacked. This makes it harder for predators to pull them out.

1 Scientists have noticed that worms usually move upwards through the soil in wet weather and downwards in hot weather. Suggest why they behave in this way.

2 Suggest how you might be able to guess the number of worms under your lawn without digging them all up.

3 Write a short story about a day in the life of an earthworm. Try to make your story exciting.

7 Light

Light is very important to life on Earth. Plants need light to make their food. Without light there would be no food for animals to eat and no oxygen in the air (see Chapter 2). We also need light to see things. You will learn a bit more about this later. Light can help us to communicate, for example traffic lights tell road users when to stop and go. Light can also be fun, for example in a firework display.

Did you know?

In the sixteenth century it was thought that Britain might be invaded by ships from France or Spain. Villages in the south of England that were near a hill or cliff often had beacons. These were iron baskets that could hold a fire.

If ships were spotted, the people would light their beacon. This would quickly be seen by people in the next village. They would then light their beacon and so on. In this way signals could be sent very quickly across the country.

There are hills in many places called Beacon Hill. Can you find any other place names that include the word beacon?

■ Beacons were the quickest way to warn people of danger

➲ Where does light come from?

Most of our light comes from the Sun. Light also comes from other sources, such as lamps, torches, candles, televisions and computer screens. Objects that give out light are described as **luminous**.

Some other objects seem to give out light but do not really do so. Shiny objects, such as mirrors, bounce light rays off their surfaces. We say that they **reflect** light. Many road signs also reflect light so well that they seem to be luminous. Surfaces that reflect light well are described as **reflective**.

■ Luminous objects give out light

Did you know?

Have you ever been outside on a moonlit night? If there is a full Moon it can be really bright. The Moon is made of rock. It cannot make its own light, so it is not luminous. It seems to be luminous because it reflects rays of sunlight off its surface and down to Earth.

Activity – luminous objects

Look around your classroom. Make a list of all the luminous objects you can see.

Now think about your home. What luminous objects can you think of there? Add these to your list.

Compare your list with the one your partner has made. Do you both have the same objects? Do you think you have identified all the objects correctly?

See if you can think up a way of checking whether an object is luminous or reflective. Describe your idea to your group. Listen to their ideas. Would they all work? Maybe you will be able to try them out.

1 Why is light important to plants?

2 Where does most of our light come from?

3 What does the word 'luminous' mean?

4 Is the Moon luminous? Explain your answer.

5 Look carefully at these pictures of objects. Write down whether each one is luminous or reflective.

A. Star

B. Television

C. Mirror

D. Diamond ring

E. Fluorescent vest

F. Candle

I. Firework

G. Moon

J. Compact disc

H. Desk lamp

Use the following words to help you to fill in the gaps in these sentences.

food   luminous   Moon   reflective   see   Sun

1 We need light to help us to _____.

2 Plants use light to help them to make _____.

3 Objects that give out light are _____.

4 _____ objects bounce rays of light from their surfaces.

5 The _____ is a luminous object but the _____ reflects light to Earth.

Exercise 7.1c extension

Peter says, 'Without light we would all be dead.'

Is he right? Explain your answer.

Did you know?

Lighthouses used to be a life-saving form of communication. Sailors could tell where they were if they could see a lighthouse because each one has its own special pattern of flashes. They could then work out where they should go to make sure they did not crash into rocks.

Now sailors use the Global Positioning System (GPS), which uses satellites in space to help you work out where you are, and other modern devices to help them to travel safely. However, lighthouses are still a useful signal. If you have a lighthouse near you, watch its flashing pattern. See if you can find the patterns for other lighthouses in the area.

If you do not have a lighthouse near you, try using Morse code. Morse code is a way of sending more complicated messages using flashes of light. It uses long and short flashes in a special pattern to represent the letters to spell a word. Find out about the Morse code alphabet and see if you can send a short message to your partner. For example, if people need help they might send a message saying 'SOS' (short for 'Save Our Souls'). How would this message be sent in Morse Code?

➲ Light travelling helps us to see

Light travels very, very fast. It travels at about 3 million metres per second. It is the fastest traveller in the Universe. Light rays always travel in straight lines from a source. If they hit a reflective surface, such as a mirror, they will be reflected in a new direction, but they will always carry on travelling in a straight line afterwards.

When you draw diagrams to show how light is travelling, always use a sharp pencil and a ruler. You should show the path of the light with a straight line. You can then add an arrow to show in which direction the light is travelling.

■ Light rays are drawn as a straight line with an arrow in the middle.

Did you know?

The Sun is about 150 million kilometres from the Earth. Light from the Sun takes just 8½ minutes to reach Earth because it travels so fast.

When light travels it can carry information. We have already seen that light can be used to send signals. It can also carry information about where it has come from. For example, a yellow lamp sends out rays of yellow light, but the light coming from a red lamp is red.

Our eyes are very clever light sensors. When light from a luminous object enters an eye, it travels to the back of the eye. The back surface of the eye is covered with millions of special light-sensing cells.

When light falls on these cells they send a message to the brain telling it all the information that came in with the light. If no light enters our eyes we see nothing because no information is entering our eyes. This is why we cannot see in the dark.

We show how light from luminous objects helps us to see them with a diagram like this.

■ The ray of red light from the traffic light tells the cyclist to stop

Taking care of our eyes

The special cells at the back of the eye can be easily damaged if too much light falls on them. We need to be careful never to look directly at very bright lights.

We must be really careful with sunlight because it is so bright and contains harmful rays that we cannot see. It is always a good idea to wear sunglasses or a cap to shield your eyes on a sunny day.

You must NEVER look directly at the Sun, even when wearing sunglasses. The light is so bright that it could seriously damage your eyes. It would be even worse to look at the Sun through a telescope or binoculars. These make the rays of light even stronger and could do even more damage.

■ Sunglasses help to protect our eyes but you should never look directly at the Sun

Did you know?

Sometimes the Moon passes between the Earth and the Sun. When this happens it covers up part or all of the Sun for a short time. We call this a solar eclipse. It is interesting to watch this happening but, of course, you should never look directly at the Sun, even when it is covered up by the Moon. To watch a solar eclipse you can use special glasses, which have a very strong filter in them. Ordinary sunglasses are not strong enough to protect your eyes.

Exercise 7.2a

1 Copy and complete this sentence: Light always travels in _____ lines.

2 Draw a simple diagram to show how a light ray helps a boy to see the flame of a candle.

3 Give two ways in which we can protect our eyes on a sunny day.

4 Why must we never look directly at the Sun?

Exercise 7.2b

Use the following words to help you to fill in the gaps in these sentences.

cap eyes straight Sun sunglasses

1 Light always travels in _____ lines.

2 We see luminous objects when light from them enters our _____.

3 To protect our eyes on a sunny day we can wear _____ and a _____.

4 We should never look directly at the _____ as it can hurt our eyes.

➡ Shadows

A **shadow** is a patch of darkness that is formed when an object blocks the light.

■ A shadow is formed when an object blocks the light

In the picture the man is standing in the sunshine. Some of the rays of light are shining onto his back. They are blocked because light cannot travel through his body. The rays of light that pass by his body are shining onto the ground. The shadow is a man-shaped hole in the light.

Explore shadows

You will need:

- a lamp
- a screen to shine the light onto (a white wall works well)
- some objects made from different materials
- some card and scissors
- a pencil and some coloured pencils
- some sticky tape
- a dark room.

First, set up your lamp so that it is shining onto the screen.

Next, try placing the objects between the lamp and the screen. Look carefully at the shadows.

Discuss your observations with your partner. Which objects make the clearest shadows? What do you notice about the shape of the shadows?

Next draw a picture of a person on the piece of card. Make it simple and about 10 cm tall. Give your person a face and some colourful clothes.

Carefully cut out your person shape and stick it to a pencil with sticky tape to make a shadow puppet.

Working Scientifically

Hold your shadow puppet between the lamp and the screen. Look carefully at the shadow.

Can you see the person's face in the shadow? Can you see what colour the clothes are? See if you can explain your observations.

If you put your shadow puppet right up against the screen, how big is the shadow?

Move the shadow puppet slowly away from the screen towards the light. How does the shadow change?

 Go further

You can see how the shadow changes as you move the puppet. To make this more scientific, you could measure the shadow.

Place a metre rule between the lamp and the screen so one end is touching the screen. Make sure that you do not move the screen or lamp while you do this investigation.

Hold the shadow puppet 10 cm from the screen and measure the shadow, from the very top of the head to the very bottom of the feet.

Now move the puppet to 20 cm from the screen and measure the shadow again. Make sure you measure from the same place each time.

Repeat this with the puppet 30 cm, 40 cm and 50 cm from the screen.

Record your measurements in a table like this:

Distance from the screen, in cm	Size of shadow, in cm
10	
20	
30	
40	
50	

Can you see a pattern in the results? Your teacher may help you to draw a line graph to show the pattern more clearly.

How clear is the pattern in your results? If it is not clear, can you suggest any way that you might improve your investigation to make your results clearer?

Shadows are always the same shape as the object that is blocking the light but not always the same size. In your investigation you will have seen that the shadow changed size as the puppet moved closer to or further away from the screen. Scientists talk about patterns like this in a special way. They would say:

The further the object is from the screen, the bigger the shadow.

Can you use your results to help you to complete this pattern sentence?

The further the object is from the light source, the _____ the shadow.

When there is a clear pattern in our results, we can use them to predict what would happen in different situations. Can you think of a way to help you to guess how big the shadow would be if you placed the puppet 25 cm from the screen?

➲ Transparent, translucent and opaque

Light can travel through some materials. Other materials do not let the light through. When you made shadows with the objects in the activity above you will have found that some of them made dark shadows and others hardly made a shadow at all.

Some materials, such as glass, allow almost all the light to pass through them. We can see clearly through them. We call these materials **transparent**. Transparent materials are useful when you need to see through them, for example in a window. Transparent materials hardly make a shadow at all.

■ Most windows are made of transparent glass because we can see clearly through it

Some materials allow some of the light to pass through but some will be blocked or scattered. We cannot see clearly through them. We call these materials **translucent**. Translucent materials can be useful if we need to let light in but we do not want anyone to see clearly, for example in the window of a bathroom. Translucent materials do not make good shadows because most of the light passes through them.

■ Frosted glass scatters some light so we cannot see clearly through it. It is translucent

Many materials block the light. No light can pass through them. We cannot see through them at all. We call these materials **opaque**. Opaque materials block all the light so they make strong, dark shadows.

■ The opaque curtains block the light

Curtains

Claire's baby keeps waking up too early. The sunlight coming into the bedroom wakes him up. Claire wants to make some curtains that will block out the light. You are going to test some fabrics to see which one will be the best to help the baby to stay asleep.

You will need:

- some fabric samples
- a torch or lamp
- a dark room
- a piece of card with a window shape cut from it
- paper clips.

To make sure that you get the right answer, you will need to make your investigation a fair test.

To make it a fair test you must change just one thing, the type of fabric you are using. Everything else must stay the same.

Discuss with your partner or group how you can use the things in the list above to test the different fabrics. Think about how you can set up your investigation so that you make sure that you are carrying out a fair test.

How will you tell which fabric is the best at blocking the light? It is not easy to be exact about how much light is coming through your fabrics. Your teacher may be able to give you a datalogger with a light sensor on it so that you can measure the light coming through. If not, you will need to look carefully and decide for yourself which fabric blocks the light best.

Think about how you can tell other people about your investigation. How will you record your results? What is the best way to show how you did the experiment? Remember that you need to say how you made it a fair test.

 Exercise 7.3a

1 How is a shadow formed?

2 What is the difference between a transparent material and a translucent material?

3 What word is used to describe a material that does not let any light through?

4 How can a shadow be made larger or smaller?

Exercise 7.3b

Use the following words to help you to fill in the gaps in these sentences.

blocks   opaque   shadow   smaller   translucent   transparent

1 A _____ is formed when an object _____ the light.

2 A material that is _____ lets nearly all the light through it.

3 A material that blocks the light is _____.

4 A material that lets some light through but that we cannot see through clearly is _____.

5 The size of a shadow gets _____ if the object is moved further away from the light source.

Some children wanted to do some experiments in the dark. They decided to make a light-proof tent to work in. Their teacher found four different fabrics.

To test each fabric they took a torch and put more and more layers of the fabric over it until no more light shone through.

Here are their results:

Fabric	Layers needed to block the light
A	4
B	4
C	5
D	2

1 Draw a bar chart to show the results of their experiment.

2 a Which fabric would you advise them to use for their tent?

  b Explain how their results show that this is the best fabric.

➲ Mirrors and reflection

At the start of this chapter we learnt that some surfaces bounce light off them. We call this **reflection**.

Some surfaces reflect light very well. Shiny surfaces such as polished metals and mirrors reflect light so well that we can see an almost perfect picture of ourselves when we look in them.

■ We can see an almost perfect picture of ourselves when we look in a mirror

Activity – investigating reflection

1: one mirror

You will need:

- a mirror
- some paper
- pencils or crayons.

Start by looking at the reflection of your own face in a flat mirror. The proper name for a flat mirror is a **plane** mirror. Do you think that it shows you exactly what you look like? Can you think of any way in which what you can see is different from the real you?

Take a piece of paper and a pencil. Write the letters OXO on the paper. Hold your mirror in different places so you can see the letters reflected in the mirror. What can you see? Can you make the letters look exactly the same in the mirror as they do on the paper? What patterns can you make?

Now write the letters TUMMY on the page and experiment with your mirror again. Do you notice anything different this time?

The picture that you can see in a mirror is exactly the same as the original shape apart from one difference. The reflected picture is reversed. When you look in the mirror you see a back-to-front picture of yourself.

Some things look the same whichever way round they are. The letters O and X look the same whether they are the right way round, back-to-front or upside-down. The letters T, U, M and Y look the same back-to-front but not if they are upside-down.

Experiment with other letters. Which ones look the same in the mirror and which ones look different when they are reflected?

See if you can write your name so that it looks correct in the mirror.

How could you use your mirror to see round a corner?

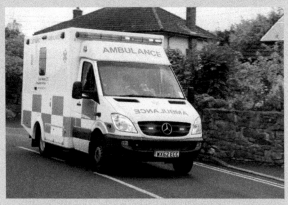

Can you suggest why the text on this ambulance's bonnet is back to front?

2: bendy mirrors

You will need:

- a flexible mirror or sheet of mirror card
- a dish-shaped mirror or shiny spoon.

Mirrors can be great fun. Have you ever visited a Hall of Mirrors at a fairground? The mirrors are all different shapes and make some very strange reflections. Use your flexible (bendy) mirror or mirror card to see what happens to your reflection when you change the shape of the mirror. Do not try this with a glass mirror. You might break it and cut yourself.

Your teacher may let you use a special dish-shaped mirror that is shiny on both sides. A shiny spoon is similar. Look at how the different shapes change the reflection.

3: two mirrors

You will need:

- two mirrors
- a piece of paper
- some tape
- crayons.

Tape the two mirrors together to make a shape like a book with a shiny inside.

Use the crayons to draw some colourful patterns on the paper.

Stand the mirrors on the paper so you can see the reflection of your patterns. Look carefully at the reflections. What do you notice about them?

Now try opening and closing the mirror 'book'. What difference does this make to the reflections?

Hold the mirrors safely and turn the paper round and round. What happens to the reflections now?

Did you know?

Have you ever looked into a kaleidoscope? A kaleidoscope is a toy that contains two or three mirrors inside a tube. Also inside the tube are some pieces of coloured translucent glass or plastic. When you look down the tube you see lots of reflections and some lovely patterns. It works in the same way as the activity above.

Kaleidoscopes are thought to have been invented in about 1815 by a scientist called Sir David Brewster.

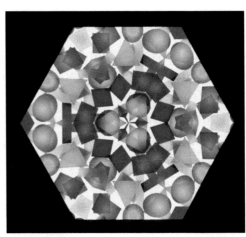

■ A kaleidoscope makes different patterns using two or three mirrors

Mirrors are very useful objects. Can you think of any more ways in which mirrors are used?

Exercise 7.4

Find words in the text to fill in the gaps in the passage below.

Light can bounce back off some surfaces. We call this _____.
Surfaces that are _____ are very good at bouncing light.

When we look in a mirror we see a picture that is nearly the same as our real face, but it is _____.

Friction and movemen

➲ Forces in action

Take a look around the room. All round you there are things moving. Your teacher may be walking around the room, the hands on the clock may be moving and the leaves on the tree outside may be blowing in the wind. We see things moving all the time but have you ever stopped to wonder why?

Things can only move if something pushes or pulls on them. Your teacher's muscles pull on the bones in his or her legs, the clock mechanism pushes the

hands round and the wind pushes against the leaves on the trees. Scientists call these pushes and pulls **forces**.

Take a look at this picture. How many things in the picture are moving? Discuss with your partner what is making the push or pull to make them move.

Forces can make things begin to move. For example, the trailer in the picture will not

move until the tractor begins to pull it. Forces can also change the speed of moving things. For example, a car will travel faster if the engine makes a bigger force to push it along. Forces can also change the shape of objects. An artist working with clay will push and pull at the clay to make a model.

⊙ Friction

Friction is one kind of force. Friction is a force that happens when two moving surfaces rub together. Wherever there is movement, there is also friction. Friction slows down moving objects. Friction may prevent objects from starting to move.

Rough or smooth?

Different surfaces make different amounts of friction. The more friction there is, the more it slows a moving object. What kinds of surface do you think make a lot of friction? How can we find out?

Comparing surfaces

You will need:

- a variety of different materials, such as carpet, smooth plastic, sandpaper, corrugated card
- a toy car.

Try rolling the car across each of the surfaces in turn. Discuss your observations with your partner or group. What happens to the speed of the car on each surface? Which surface do you think has the most effect on the movement of the car?

Do you think that you have just carried out a fair test? Think about how you made the car move. Were you able to make it move in exactly the same way each time?

Work with your partner or group to plan a way to roll the car across the surfaces in a fair test. What will you keep

Working Scientifically

the same? What will you measure to get your results? How will you record your results? How will you know which surface made the most friction?

When you have carried out your investigation, look at the surfaces again. Can you say what kind of surface makes the most friction? See if you can explain why this happens.

Draw a neat, labelled diagram to show how you carried out your experiment. Discuss with your group whether you think there is any way you might have changed your investigation to make it better.

In your investigation, you should have found that the rougher surfaces caused the most friction. Smooth surfaces make much less friction.

When two rough surfaces rub together, the bumps and dips on the surfaces get caught up with one another, resisting movement. Smooth surfaces slide past each other easily and the friction force is much less.

Useful friction

Friction can be really useful. Friction is sometimes known as **grip**. When you walk across the floor it is useful to have some friction between your shoes and the floor surface. This stops you from sliding around and falling over. Cars need friction between the tyres and the road surface to stop them skidding. If you look carefully at the soles of your shoes, you will see that they are not smooth. The pattern on them helps them to grip the floor. The tread pattern on the tyres of a car does the same job. Can you think of any other situations where friction is useful?

■ The tread on the tyre makes a lot of friction to help the tractor grip the road

Exercise 8.1a

1 What is a force?

2 What effects do forces have when they act on objects?

3 How is friction caused?

4 Which type of surface causes the most friction?

5 Explain in your own words why these surfaces cause so much friction.

6 In icy weather we often spread grit on the roads. Suggest how this helps motorists to stay safe.

Exercise 8.1b

Use the following words to help you to fill in the gaps in these sentences.

increases   pull   push   rough   rub   skidding   shape   speed

1 A force is a _____ or a _____.

2 Forces can change the _____ or _____ of an object.

3 Friction is made when surfaces _____ together.

4 Surfaces that are _____ make the most friction.

5 The tread on a car tyre _____ the friction between the tyre and the road to stop the car from _____.

Comparing shoes

You will need:

- several shoes with different grip patterns on them
- a plank of wood
- a metre ruler
- a calculator.

To make this a fair test, try to choose shoes that are all about the same size and all weigh about the same. If they are very different you could put some weights inside the lighter ones to even them out.

Working Scientifically

Look at the treads on the shoes. Predict which shoe you think will have the best grip and which will have the worst. Give reasons for your prediction.

Place one shoe on the plank of wood. Slowly lift one end of the plank until the shoe just begins to move.

Use the metre ruler to measure how high you have lifted the plank when the shoe begins to move. Record this result in a table like this.

Shoe	Height of end of plank, in cm

Repeat this with each of the shoes in turn.

Draw a bar chart to show the height of the plank for each shoe.

How do the results tell you which shoe had the best grip? Can you say which had the worst grip? Do these results match your original prediction?

Now put the plank down, put the first shoe back on and test it again. Do you get the same answer as you did before? What happens when you test it for a third time?

Often in science we do not get the same answer each time we carry out an experiment. When this happens we say that the results are not very **reliable**. How reliable do you think your results are? Scientists usually do their experiments more than once to check how reliable the results are. If they are always the same, they are reliable.

If you have time you could check to see if your results are reliable.

Test each shoe three times and then record the results for each test in a table like the one below.

	Height of end of plank, in cm			
Shoe	1st test	2nd test	3rd test	Mean

Are the results the same each time? Would you say that your results are reliable?

 ## Go further

You now have three results for each shoe instead of one, so which result is the right one? The answer is that we cannot say which one is right. When scientists have this problem they do a special calculation that turns their three results into a single one.

You could do this calculation for your results.

Use a calculator to add together the three results for your first shoe.

Remember to press the = key to get the total.

Now divide this total by 3.

This gives you a single result that should be similar to the three results you measured. This is what we call the mean or average of your results.

Write this value into the last column in the table.

Repeat this for the other shoes.

Friction is useful in another, very important way. The friction force always works against the force that is making something move. This means that it will slow things down. The brakes on a car or bicycle work because the friction between the brake pads and the wheel make the wheel slow down and then stop.

■ Friction between the brake pad and the wheel makes the bike stop

Friction can be a nuisance

Friction is not always useful. Try rubbing your hands together quite hard. What do you feel? When surfaces rub together they often cause heat. They can also rub bits off each other causing the surfaces to become worn. If we want things to move smoothly past each other we do not want friction to slow them down and wear them out. We need to make these surfaces as smooth as possible. Sometimes we add oil or grease to moving surfaces to help them move past each other more easily.

■ Oil helps to reduce friction between moving surfaces

When things move through the air, they rub against the tiny air particles, making friction. This is called **air resistance**. If you try running across the playground holding a big sheet of card in front of you, you can feel the air pushing against the card, making it hard to run. Air resistance, like all friction, slows the movement of objects.

Did you know?

When rocks from space enter the Earth's atmosphere they are moving very fast. As they rub against the air particles the friction (air resistance) heats them up so much that they begin to burn up. We see these as shooting stars or meteors.

If people want to move very fast through the air or water, they need to reduce the effect of air resistance. They can do this by making it possible for the air to move past them smoothly. We call this **streamlining**.

One example of streamlining in action is seen in bicycle racing. The cyclists in the photograph want to travel as fast as possible. Their clothes and bicycles are all specially designed to make them streamlined. The wheels of the bikes are solid because air catches in the spokes of normal bicycle wheels and this would slow them down. Their helmets allow the air to pass smoothly across them. The cyclists bend down over the low handlebars so that bodies do not catch the air. Their clothes are smooth and fit tightly so they do not flap around in the wind.

Bloodhound SSC is a car that is being built to break the land speed record. If it is to reach the required speed of about 1690 km/h (1050 miles/h), it needs to be as streamlined as possible. Look carefully at its shape. Can you think of any other things that are streamlined like this?

■ Bloodhound SSC is streamlined

Activity – Bloodhound SSC

Use the internet to find out more about the Bloodhound SSC project. You could find out about the engines that make the forces to push it so fast along the ground. You could also find out about the other forces acting on the car as it rushes along.

Did you know?

Bloodhound SSC is a very modern car. Streamlining is a trick that we have learnt from nature. Many animals have streamlined shapes to help them to move smoothly through the air or water. Think about the shape of a dolphin. It has a smooth body and a pointed nose, just like Bloodhound SSC. A gannet tucks its wings in so it looks like a fighter jet to help it to dive into the water to catch a fish without hurting itself.

■ Dolphins' streamlined shape helps them to move fast through the water

Exercise 8.2a

1 Give three examples where friction is useful.

2 Give three examples where friction is a nuisance.

3 Explain why you might put oil on a door hinge.

4 What name is given to the friction force made when things move through the air?

5 What is meant by the term 'streamlined'?

6 Suggest why a submarine has a similar shape to a whale.

Exercise 8.2b

Use the following words to help you to fill in the gaps in these sentences.

grease   grip   nuisance   oil   smooth   streamlined   useful

1 Another name for friction is _____ .

2 Friction can be _____ or a _____ .

3 We can reduce friction by making surfaces very _____ .

4 To make moving surfaces more slippery we can add _____ or _____ .

5 A _____ shape moves more smoothly through air or water.

Exercise 8.2c: extension

Sharks have narrow bodies, smooth skins and pointed noses. Puffer fish are rounded and have rough, spiny skins. What do you think this tells us about the behaviour of these two fish?

Measuring forces

Scientists always like to measure things. To measure something you need a piece of apparatus to measure with. You also need units to measure in. If we wanted to measure the length of a running track, we would use a tape measure and measure the distance in metres. To measure the temperature of a cup of tea we would use a thermometer and measure in degrees Celsius.

Forces are measured in units called **newtons (N)**. They are named after a famous English scientist called Sir Isaac Newton. The measuring device for forces is called a **force meter**.

We can use a force meter to measure the force needed to move something.

Activity – moving heavy objects easily

You will need:

- a brick
- string
- some marbles
- a tray to keep the marbles from rolling away
- some pieces of dowelling or round pencils
- a force meter.

Tie a piece of string round the brick. Place the brick in the centre of the tray with a flat side downwards and hook the force meter into the string.

Pull gently on the force meter until the brick just starts to move. Look at the force reading on the force meter when the brick just starts to move. Write this down. Remember to write the units.

Now try placing some marbles under the brick and try pulling again. What happens to the force needed to move the brick?

Repeat the experiment but this time put the dowelling or pencils under the brick.

Can you explain how the marbles and the dowelling or pencils make it easier to pull the brick?

People have needed to move heavy things for thousands of years. We have machines to help us now, but ancient people had to use some clever tricks to help them. Here is the story of how some ancient people used science and the materials around them to move some very heavy pieces of rock.

● Ancient builders

Over four thousand years ago in Wiltshire in the south of England people started to build Stonehenge. Stonehenge is a circle of massive stone arches, and an inner circle of smaller stones. No one knows exactly why it was built. It may have been a kind of temple, or a huge sundial, or a place to observe the movement of the stars and planets.

■ Stonehenge

The stones of the inner circle may not be the largest, but they still weigh about 4 tonnes each. They came from the Preseli mountains in Wales, about 240 miles away. Moving such huge stones over this distance would have been a very difficult job.

We do not know for certain how the stones were moved. It is likely that for part of the journey they would have been put on barges and floated along a river. Even so, they would have been moved over land for long distances. The wheel had not been invented by the time Stonehenge was built. Many historians believe the stones were placed on rollers, probably whole tree trunks, and pulled along using leather ropes. As the stones moved across the rollers, the roller at the back would roll out from under the stone. Men would carry it to the front of the stone and then pull the stone onto it, releasing the next one from the back. What a lot of work!

The outer circle is made of even bigger stones, up to 50 tonnes each. These came from much nearer the site but they are far too heavy to have been taken by river. They would have sunk the boat! These ones must have come over land. Some people have calculated that it would have taken 500 men pulling on leather ropes to move each stone and another 100 men to move the rollers!

■ Over 500 men would have been needed to move each stone at Stonehenge

Somehow these massive stones were tipped up on end and settled into pits in the ground to make the stone circle. Large stones, or lintels, were then heaved up and placed across the top. Stonehenge was built thousands of years ago but the ancient people who built it obviously knew a lot about forces!

Exercise 8.3

1 Where is Stonehenge?

2 When did the building of Stonehenge begin?

3 How far did the stones for the inner ring have to be transported?

4 How heavy are the largest stones in the outer ring of Stonehenge?

5 Explain, using your knowledge of forces, why the builders of Stonehenge would have found it difficult to move these stones.

6 Describe in your own words how these huge stones might have been transported.

7 Explain why this method would make it easier to move the stones.

8 What other method was probably used for moving some of the stones?

9 Why was it not possible to use this method for all the stones?

10 What do historians believe Stonehenge might have been used for?

Magnets

Magnetic forces

In the last chapter you learnt that a force is a push or a pull. These forces act on objects. Forces can change the way an object is moving or change its shape.

To make a football move across the field, you need to kick it. To open a door you need to pull it. A boat will move when the wind blows onto its sail, pushing it along. Making a force on an object usually needs contact between the object and whatever is making the force.

There is one type of force that can work without touching the object. Magnets are able to pull or push on some objects without touching them.

Activity – magnetic pulls

You will need:

- a bar magnet
- a paper clip.

Place the paper clip on the table. Place the magnet on the table a little way away from the paper clip and slide it gently towards the clip.

When the magnet gets close enough to pull on the clip, the clip hops across the table to join onto the magnet. The magnetic force has pulled on the paper clip.

Turn your magnet round and try again. Does it make any difference which end of the magnet approaches the paper clip? What happens if you slide the magnet sideways on?

All bar magnets have two 'ends' called **poles**. We will learn more about these later in the chapter. They are the strongest parts of the magnet. The middle part is not so strong so it is not as good at pulling the paper clip.

⊙ Magnetic or non-magnetic?

Paper clips are made from a metal called **steel**. Steel is attracted towards magnets. We call it a **magnetic material**. Are all materials magnetic? Take your magnet round the classroom and see what it will stick to. Do not put it near the computer or interactive whiteboard as it could cause damage. Can you see any pattern in your observations?

These children have some ideas about magnetic materials. Discuss their ideas with your partner or group. Do you think they are right? Can you think of ways to test their ideas to see who is right?

I think that the magnet will stick to anything made of metal.

I think it will only stick to items made from iron or steel.

I think that it will stick better to some items than others.

I think it will stick to anything that is silver coloured.

Magnetic materials

You will need:

- a bar magnet
- a variety of objects made from different materials, such as steel, copper, iron, plastic, wood, fabric, cooking foil.

Use the above items to find out which materials are magnetic and which are not. Record your results in a neat table.

Look back at the children's ideas in the picture. Can you now say which of the ideas is most accurate?

When you did the activity above, you will have found that the magnet stuck to some materials and not to others. Some materials are magnetic but most are **non-magnetic**.

Often people think that magnets will stick to all metals but you have found that this is not true. Most of the magnetic objects that you found in the classroom were probably made from steel. Steel is very useful. It is strong and can easily be made into lots of different shapes. Steel is magnetic because it contains another metal called **iron**. Iron is the only common magnetic metal.

Did you know?

In 1992, the material used to make 1p and 2p coins was changed. Coins made before 1992 are made from bronze, which is a non-magnetic metal. Coins made since that date are made from steel, covered with a thin layer of copper, and they are magnetic.

Exercise 9.1a

Use words from the text and the results of your experiments to help you to fill in the gaps in these sentences.

1 Objects made from _____ or _____ will stick to a magnet.

2 A material that sticks to a magnet is said to be _____.

3 Two examples of materials that will not stick to a magnet are _____ and _____.

4 A material that does not stick to a magnet is said to be _____.

Exercise 9.1b: extension

1 Alex wants to make a magnetic fishing game. She will make some card fishes and try to catch them using a small magnet on a string. How could she make her card fishes magnetic so she can catch them?

2 Fizzy drinks often come in cans. Some cans are made from steel and some are made from another metal called aluminium. Design a machine that uses a magnet to separate the steel cans from the aluminium ones so they can all be recycled.

➲ Magnetic force fields

How can a magnet pull on an object without touching it? Magnets are surrounded by an invisible force field. When a magnetic object comes into the force field it is pulled towards the magnet. We say that it has been **attracted** by the magnetic field.

Activity – seeing the invisible

It is possible to make the force field visible using tiny pieces of iron called **iron filings**.

You will need:

- a bar magnet
- iron filings in a sprinkler pot
- a large sheet of paper.

Put the magnet on the table and lay the sheet of paper over the top.

Carefully sprinkle the iron filings over the sheet of paper and all round the magnet. Do not use too many iron filings. Just sprinkle enough to see the pattern clearly. The iron filings line up with the force field around the magnet so that you can see the shape of it.

When you have finished, carefully lift the paper and pour the iron filings back into the pot. Take care not to get any of them on the magnet as they are hard to clean off again. Wash your hands after this activity and take care not to get the iron filings in your eyes.

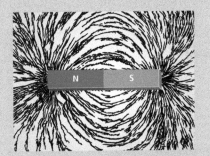

A force field demonstrator uses iron filings to show the force field around a bar magnet

You may also be able to use a force field demonstrator like the one in the picture. These contain iron filings sealed inside a frame. Place the demonstrator over the magnet and tap it gently. The iron filings will line up with the force field.

The shape of the force field is the same around all bar magnets. You can see how the lines cluster round the poles (ends) of the magnet. You will remember that these are the strongest parts of the magnet. Using these methods to see the force field makes it seem as if it is flat. However, the field is really all round the magnet and works in all directions. If you repeat the experiment with the iron filings, but this time pile lots of the filings around the magnet, you may be able to see how they make a 3D shape.

If you have some different shaped magnets, for example a horseshoe magnet, a ring magnet and a button magnet, you could compare the shapes of their force fields using the iron filings or force field demonstrator. See if you can work out where the poles are on these magnets.

Magnetic force fields are very useful. We use them for all sorts of things. Maybe you have played with a train set or construction toy that uses magnets to stick the parts together. The door of the fridge in your kitchen almost certainly has a magnet in it to keep it closed. Most motors also contain magnets. Can you think of any other places where magnets are used?

⊃ More about poles

As you have already learnt, a bar magnet has two ends called 'poles'. It may surprise you to learn that the whole Earth is like a huge bar magnet. It also has an invisible magnetic force field and two poles. The Earth's two poles are called the magnetic north pole and the magnetic south pole.

Activity – magnetic poles

You will need:

- a bar magnet
- a piece of string.

Carefully tie the string round the middle of the bar magnet and adjust the position until it hangs level.

Hold the end of the string and allow the magnet to come to rest. Notice which way it is pointing.

Give the magnet a little push and then let it come to rest again.

What do you notice about the way it is pointing? Try several times. Does it do the same each time?

When we hang a bar magnet on a piece of string it will always swing around until it comes to rest lined up with the lines of the Earth's magnetic field. One end of the magnet will point toward the Earth's north pole. We call this end of the magnet the **north-seeking pole**. This is often marked with a letter N on the magnet. The north-seeking pole of the magnet is also often coloured red. The other end, which points towards the Earth's south pole, is called the **south-seeking pole**. This is marked with a letter S on many magnets.

Why does a magnet not spin round to line up with the Earth's magnetic force field when it is lying on the table? (Hint: think about what you have learnt in Chapter 8.)

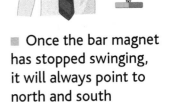

■ Once the bar magnet has stopped swinging, it will always point to north and south

This ability of magnets to line themselves up with the Earth's magnetic force field is very useful. Have you ever used a **compass** to help you to work out in which direction you are facing? A compass contains a long, thin magnet, called the **compass needle**, which is supported on a very tiny pin. This allows it to swing round to line up with the Earth's magnetic poles. This tells us where north and south are. We can then use a map to help us work out where we are facing or which direction we need to travel to get to our destination.

■ A compass is very useful when we are map reading

Did you know?

The Earth's magnetic force field is very important to life on Earth. It acts as a shield to stop harmful particles from the Sun coming to the surface of the Earth. We can sometimes see this happening because the particles can glow different colours as they meet the Earth's magnetic field. In the Northern hemisphere we call this glow the 'aurora' or 'Northern Lights'.

■ The Northern Lights are caused when particles from the Sun meet the Earth's magnetic field

Two magnets

You have already learnt that a magnet can attract anything made from a material that is magnetic. Let's see what happens when two magnetic force fields come together.

Activity – attracting and repelling

You will need:

● two bar magnets.

First, work out which is the north-seeking pole and which is the south-seeking pole of each magnet.

Take the two magnets and place them on the table, a little way apart, with the north-seeking pole of one facing towards the south-seeking pole of the other.

What happens when you push the magnets closer together?

When two different poles (known as **opposite** poles) come together, the two force fields attract each other and the magnets will stick to each other.

Now turn one magnet round so that the two north-seeking poles are facing each other.

What happens this time? If you pick up the magnets and try to push the two north-seeking poles together you will be able to feel the two force fields pushing each other away. We say that they **repel** each other. Try with the two south-seeking poles. If you have two poles the same (we call them **like** poles) they will always repel each other.

Did you know?

Magnets can attract or repel each other. They can attract objects made from magnetic materials but they cannot repel anything other than another magnet. This is because the push is caused by the two force fields of the magnets pushing against each other.

Magnets come in all shapes and sizes. Here is one called a ring magnet. Its poles are on the top and the bottom of the ring.

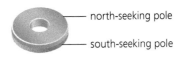

north-seeking pole
south-seeking pole

■ A ring magnet

What is happening in this picture? A series of ring magnets has been threaded onto a stick. The magnets seem to be floating. Use your knowledge of magnetic poles to explain why.

Exercise 9.2a

Each of the following sentences has something wrong with it. Read them carefully and then rewrite each one, changing one or two words to make it correct.

1 Two south-seeking poles of a magnet will attract each other.

2 A north-seeking pole will repel a south-seeking pole.

3 If you hang a bar magnet up it will point to east and west.

4 Plastic is non-magnetic, so it is repelled by a magnet.

Exercise 9.2b: extension

Emily was given three identical-looking pieces of metal. She was told that one was made from steel, one was made from aluminium and one was a magnet. Describe how she could use another magnet to find out which was which.

➡ Investigating magnetic forces

Magnetic forces can act at a distance. They can also act through other materials.

Activity – magnetic forces

You will need:

- a magnet
- sheets of a variety of different materials, such as paper, card, fabric, wood, copper, steel
- a paper clip.

Place the magnet on one side of each of the materials and the paper clip on the other side. Can the magnet hold the paper clip through the material?

Magnetic forces can work through most materials. If you had some steel or iron, you will have found that the magnet sticks to the metal sheet but cannot also hold the paper clip.

Magnets can work through paper, card, wood, fabric and other materials as long as they are not too thick. If the material is thicker than the magnet's force field, the magnet cannot reach through the material. You can use this to test the strength of magnets.

Testing the strength of magnets

You will need:

- a variety of magnets
- a very thick book
- a paper clip.

Open the book and place the paper clip on page 1.

Take a magnet and place it a few pages further on in the book and see if it can attract the paper clip. If not, use fewer pages between the magnet and the paper clip. If it can hold the paper clip, add more pages. Find the maximum number of pages that the magnet can work through. (Hint: use the page numbers to tell you how many pages there are but remember that each page has two sides!) Write your results down in a neat table.

Now repeat with the other magnets. Find out how many pages each one will work through and record the results in your table.

Working Scientifically

How do your results show you which magnet is strongest? Sort the magnets into order from strongest to weakest. Draw a bar chart of your results.

This is just one way to test the strengths of your magnets. Can you think of another way to test them? Discuss it with your partner or group. If you get stuck, look back at the very first activity in this chapter. How could this activity be adapted to show the strength of the magnets?

Did you know?

Scientists sometimes need to check if there is enough paint on a structure so that the iron or steel underneath the paint is protected. They do this by using magnets. They can estimate how many layers of paint there are on structures such as steel bridges by measuring the strength of the magnetic force between a magnet and the steel under the paint. The weaker the force, the more paint there is on the surface.

A magnetic challenge

Your challenge is to design and make a simple toy or useful device that includes a magnet. Think about what you have learnt about magnetic force fields and magnetic materials. You could choose to use two magnets so that they repel each other.

Draw up a neat design. Label it clearly to show how it works. When you have finished your design, your teacher will give you one or two small magnets and some materials to make your toy or device.

Show your design and model to the rest of the class. Describe clearly how it works and how you made it. Say whether you think your design is as good as it could be or whether you would make any improvements if you made it again.

Make a display of your designs and the finished models. Try to make your display as informative as possible so other people can learn all about magnets.

Glossary

Air resistance The friction force between moving objects and the air.

Attract To pull towards (a magnet).

Backbone The group of bones that run up the back of the skeleton between the pelvis and the skull. Also called the vertebral column.

Balanced diet A diet that contains all the nutrients we need in the right quantities to keep us healthy.

Biology The study of living things.

Blood The red liquid that carries food, oxygen and other materials to the different parts of the body.

Brain The organ that controls the body.

Calcium A mineral salt needed for healthy bones.

Carbohydrates Nutrients, found in foods such as bread, pasta and potatoes, which give us energy.

Carbon dioxide A gas in the air that is used by plants when making food in the process called photosynthesis.

Carnivore An animal that eats only meat.

Carpel The female part of a plant, which produces the eggs.

Cells The microscopic building blocks that make up the bodies of all living things.

Chlorophyll The green pigment (colour), found in the leaves and stems of plants, which takes in light energy from the Sun.

Clay soil Soil that contains mostly very tiny pieces and very small air spaces, through which water drains very slowly.

Compass Device using a magnet to show the position of the Earth's magnetic north and south poles.

Compass needle The free-moving magnet inside a compass, which spins round to point to north and south.

Control Part of an experiment that is set up to provide a result to compare with others.

Crust The layer of solid rock on the surface of the Earth.

Decomposers Animals and fungi that break down the remains of dead plants and animals and mix them with the soil.

Diet All the food and drink that we take into our bodies.

Digest To break down food into a form that can be used by the body.

Environment The surroundings or conditions in which something is found.

Excretion Getting rid of waste from the body.

Exoskeleton The hard outside case surrounding the body of some invertebrates, e.g. insects.

Extinct When there are no more living members of a species.

Fats Nutrients, found in meat, fish and dairy products, needed in small quantities to store energy and keep us warm.

Fibre Material, found in fruit and vegetables, that is needed to keep food moving through the intestines properly.

Flexible Bendy.

Force A push or pull causing something to begin to move or change its movement.

Force meter A device for measuring forces.

Fossil The remains of an animal or plant that died millions of years ago, found in a rock.

Friction A force created when two surfaces rub together.

Germination When a seed begins to grow by putting out its first root.

Gravel Rocky pieces in the soil that are more than 2 mm across.

Grip Another name for friction.

Growth Getting bigger.

Heart The organ that pumps the blood around the body.

Herbivore An animal that eats only plants.

Humus The remains of dead plants and animals that have become broken down and mixed into the soil.

Intestines Part of the body where digested food is taken into the blood for transporting around the body.

Invertebrate An animal that does not have an internal skeleton/backbone.

Iron The most common magnetic metal.

Iron filings Tiny pieces of iron, used to show up a magnetic field.

Joint A place where two or more bones are joined together.

Life cycle The story of the life of a living thing from baby to adulthood.

Life processes Processes that are carried out by all living things to keep them alive and healthy.

Like (of magnetic poles) The same, i.e. both north-seeking or both south-seeking.

Loam Soil that contains a mixture of different sized pieces and lots of humus. The best soil for growing plants.

Luminous Giving out light.

Lungs The organs that take oxygen from the air into the body and remove carbon dioxide.

Magnetic material A material that is attracted towards a magnet.

Microscopic Too small to be seen without a microscope.

Mineral salts Substances taken in from the soil in very small quantities by plants to keep them healthy; nutrients, such as calcium, needed in small quantities in the diet for healthy growth.

Movement Changing position, moving all or part of the body.

Muscles Groups of special cells that can pull on bones to move the body.

Newtons Units of force.

Non-magnetic Not attracted to a magnet.

North-seeking pole The end of a magnet that turns towards the Earth's magnetic north pole.

Nutrient A substance in our food and drink that is needed to keep the body working properly.

Nutrition Feeding.

Obese Overweight.

Omnivore An animal that eats plants and meat.

Opaque Light cannot travel through.

Opposite (of magnetic poles) Different, i.e. one north-seeking and one south-seeking.

Organism A living thing.

Organs Parts of the body that have a particular job to do. Organs are made up from lots of cells grouped together.

Oxygen The gas in the air that is needed by nearly all living things to help them release energy from food in the process called respiration.

Photosynthesis The process used by plants to make their own food.

Pigment Coloured substance, e.g. the green pigment (chlorophyll) in leaves.

Plane (of a mirror) Flat.

Poles The 'ends' of a bar magnet. The strongest parts of a magnet. May be north-seeking or south-seeking.

Pollen The special male cells produced by the stamen.

Pollination The transfer of pollen from one flower to another, usually by insects on the wind.

Predator An animal that catches other animals to eat.

Proteins Nutrients, found in meat, fish, eggs and dairy products, used by the body to make new cells.

Reflect Bounce back.

Reflection The act of reflecting (bouncing back) a ray of light; the image seen in a mirror.

Reflective (of a surface) Good at reflecting light.

Reliable (of experimental results) Similar every time they are measured.

Repel To push away from (a magnet).

Reproduction Making more members of the species. Animals have babies to reproduce; most plants make seeds.

Respiration The life process that allows living things to get energy from their food.

Rib cage The group of thin bones that support the chest and protect the heart and lungs.

Sandy soil Soil that contains a lot of large pieces and big air spaces, through which water drains quickly.

Sediment A mixture of sand, mud and parts of animals and plants that settle at the bottom of a lake or sea.

Sedimentary rock A type of rock made up from layers of sediment.

Seed dispersal Spreading seeds away from the parent plant so they grow in a new place.

Seedling A tiny new plant recently grown from a seed.

Shadow An area where the light has been blocked by an opaque object.

Silt Very tiny rocky pieces in the soil that are almost too small to see.

Skeleton The bones that support and protect the body and make it possible for us to move easily.

Skull The part of the skeleton in the head that protects the brain.

South-seeking pole The end of a magnet that turns towards the Earth's magnetic south pole.

Species A particular type of living thing.

Stamen The male part of a flower that produces the pollen.

Steel A very common material containing iron, which makes it magnetic.

Stomach An organ that digests food and makes it ready to be used by the body.

Streamlining When an object or animal is shaped to reduce friction between it and the air or water that it moves through.

Translucent Some light can travel through but we cannot see clearly through.

Transparent Light can travel through and we can see through clearly.

Vertebral column The group of bones that run up the back of the skeleton, between the pelvis and the skull. Also called the backbone.

Vertebrate An animal that has an internal skeleton/backbone.

Vitamins Nutrients, such as vitamin C, needed in small quantities in the diet to prevent disease.

Volcano A mountain made when liquid rock escapes through a gap in the Earth's crust.

Weathering When rocks are broken up by the weather or in a river or stream.

Index

air
    oxygen in 6, 13, 30
    in soil 67, 69
    in water 27
air resistance 101, 119
animals
    diet/nutrition 5, 41–42
    growth 49–50
    movement 4, 48–49
    seed dispersal by 20
    skeletons 46–47
    in soil 71–75
    streamlining 102
attraction 110, 114–15, 119
average 99
backbone 43, 44, 119
bacteria 5
balanced diet 35–36, 119
beacons 76
bicycle racing 101
biology 3, 119
blood 30, 119
Bloodhound SCC 102
bones 43–45
brain 30, 31, 44, 119
brakes 100
calcium 34, 119
carbohydrates 33, 119
carbon dioxide 119
    in human body 30
    in plants 13, 14
carnivores 42, 119
carpels 18, 119
cells 5, 119
chlorophyll 9, 13, 28, 119
clay soil 69, 119
compass 113, 119
compass needles 113, 119
control 11, 119
core 51
crust 51, 119
crystals 52
curtains 88–89
decomposers 63, 119

diet 33–36, 41–42, 119
    see also nutrition
digestion 31, 119
drainage 67–68
Earth
    magnetic field 112–14
    structure 51
earthworms 47, 73–75
energy
    nutrients for 33–34
    in plants 13
environment 6, 119
excretion 6, 119
exoskeletons 47, 50, 119
extinction 5, 119
eyes 80–81
fats 33–34, 36, 119
fibre 35, 120
flexible 92, 120
flowers 9, 10, 17–18
food *see* diet; nutrition
force meters 104–5, 120
forces 94–95, 120
    magnetic 107, 116–17
    measuring 104–5
fossils 56–57, 120
friction 95–101, 120
fruits 21, 35
fungi 63
germination 24–28, 29, 120
giant pandas 5
gravel 62, 120
grip 96, 98, 120
groups 1–3, 22
growth 3, 5, 120
    animals 49–50
    nutrients for 34
    plants 11–13
heart 30, 31–32, 44, 120
herbivores 41, 120
hibernation 8
humus 63, 120
intestines 30, 31, 120
invertebrates 47, 48, 50, 120

iron 109, 120
iron filings 111, 120
joints 44, 45, 120
kaleidoscopes 93
kidneys 30, 31
leaves 9, 10
life cycles 28–29, 120
life processes 3–8, 120
light
    plants and 12–13, 14, 28
    properties 80
    sources 77
lighthouses 79
like poles 115, 120
limestone 57–58
liver 30, 31
loam 69, 120
luminous objects 77, 120
lungs 30, 44, 120
magnetic fields 110–15
magnetic materials 108–9, 120
magnets 107–18
    poles 108, 112–13,
    114–15, 121
    strength of 117–18
mantle 51
mean 99
microscopic 3, 120
mineral salts 120
    in human diet 14, 34
    in plants 14, 15, 63
mirrors 90–93
Moon 77, 81
Morse code 79
movement 3, 4, 44, 48–49, 120
muscles 43, 44, 120
nectar 18
nerves 31
newtons (N) 104, 120
non-magnetic materials 109, 120
north-seeking pole 113, 120
Northern Lights 114
nutrients 33–35, 121

nutrition 3, 5, 121
    animals 5, 41–42
    humans 31, 32–36
    plants 5, 13–14
nutritionists 37
obesity 34, 121
observations x
omnivores 42, 121
opaque materials 88, 121
opposite poles 115, 121
organisms 3, 121
organs 30–31, 44, 121
oxygen 121
    in human body 30
    in plants 13, 14, 26–28
    for respiration 6, 13
photosynthesis 13, 121
pigment 9, 121
plane mirrors 91, 121
plants
    growth 11–13
    life cycle 28–29
    movement 4
    nutrition 5, 13–14
    reproduction 17–19
    structure 9–11
    transport systems 14–16
    see also seeds
poles 108, 112–13, 114–15, 121
pollen 18, 121
pollination 18, 19, 29, 121
Pompeii 54
predators 47, 121
proteins 34, 36, 121
pulse rate 31–32
rabbits 5
records x

reflection 77, 80, 90–93, 121
reflective surfaces 77, 80, 121
reliable results 98, 121
reproduction 3, 5, 121
    in plants 17–19
repulsion 115, 121
respiration 6, 121
rib cage 43, 44, 121
ring magnets 115
rocks
    formation 51–52, 55–58
    in soil 62–63
    testing 59–60
    uses 58–59, 60
roots 9, 10, 14–15, 24
safety xi
sandy soil 69, 121
science viii
scientists viii–xi
scurvy 35, 39–40
sediment 55, 63, 121
sedimentary rocks 55–58, 121
seedlings 11–13, 121
seeds 18
    dispersal 20–23, 29, 121
    germination 24–28, 29
sensitivity 6
shadows 82–86, 121
shoes 96, 97–99
sieving soils 64–65
silt 62, 122
skeletons 31, 43–44, 46–47, 122
skull 43, 44, 122
soils
    comparing 64–68
    composition 61–63, 67–68
    life in 71–75
    naming 69

solar eclipse 81
sorting 1–3, 22
south-seeking poles 113, 122
species 5, 122
stalactites/stalagmites 58
stamens 18, 122
starch 33, 36
steel 108, 109, 122
stems 9, 10
stomach 30, 31, 122
Stonehenge 105–6
streamlining 101–2, 122
sugars
    in human body 33, 36
    in plants 13, 14
Sun 77, 80, 81
surfaces, friction and 95–96
tarantula spiders 50
temperature, plants and 12–13, 26–28
translucent materials 87, 122
transparent materials 86–87, 122
vertebral column 43, 44, 122
vertebrates 46–47, 50, 122
vitamin C 35, 39–40
vitamins 34–35, 39–40, 122
volcanoes 52–54, 122
water
    in human body 35
    in plants 12–13, 14, 15, 26–28
    seed dispersal by 21
    in soil 67
weathering 62, 122
wind
    pollination by 19
    seed dispersal by 20–21